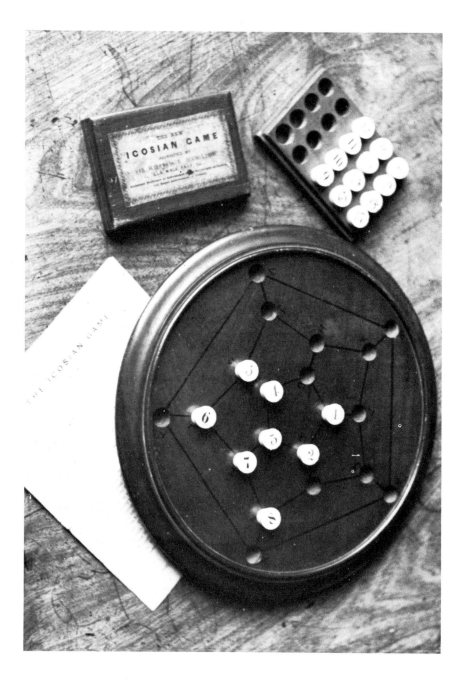

Hamilton's Icosian Game (Chap. 2)

Graph Theory

1736-1936

NORMAN L. BIGGS

E. KEITH LLOYD

ROBIN J. WILSON

CLARENDON PRESS · OXFORD

1976

Oxford University Press, Ely House, London W.1

OXFORD LONDON GLASGOW NEW YORK
TORONTO MELBOURNE WELLINGTON CAPE TOWN
IBADAN NAIROBI DAR ES SALAAM LUSAKA ADDIS ABABA
KUALA LUMPUR SINGAPORE JAKARTA HONG KONG TOKYO
DELHI BOMBAY CALCUTTA MADRAS KARACHI

ISBN 0 19 853901 0

Typeset in Northern Ireland at the Universities Press, Belfast
Printed in Great Britain by J.W. Arrowsmith Ltd., Bristol

L. EULER (1707–83)

D. KÖNIG (1884–1944)

Preface

MATHEMATICIANS have often pursued their researches in an erratic and intuitive way, rather than by the clear light of logic; consequently, the historical development of the subject frequently differs considerably from the systematic approach which one finds in most textbooks. In this book we shall follow an historical approach, and give a self-contained introduction to the subject of graph theory. We hope that the reader will thereby come to appreciate the complex web of ideas and influences which come together to form a mathematical theory.

Our decision to cover the period 1736–1936 is the result of a convenient historical accident. In 1736 the first article on a topic relating to graph theory was written by the Swiss mathematician Leonhard Euler; just two hundred years later, in 1936, the first full-length book on the subject, written by Dénes König, was published. Of course, graph theory did not stop in 1936, and we have not felt obliged to exclude all reference to later work in the subject.

The central feature of this book is a set of thirty-seven extracts taken from the original writings of mathematicians who contributed to the foundations of graph theory. Where necessary, these extracts have been translated into English, and they have been edited by the omission of certain sections, but no other significant changes have been made. A list of these extracts may be found on pages (ix)–(x).

The book has ten chapters, Each chapter deals with a particular theme in graph theory, and contains three or four main extracts; these extracts are linked by a commentary which traces the historical development of the theme. Superimposed on this structure is the framework of a conventional textbook, wherein the relevant mathematical terminology and notation are explained, in logical progression, as they are required.

There are a few conventions regarding the organization of this book which may need some explanation. The references within each chapter (including those references which occur in the extracts) are labelled in a single numerical sequence, and they are listed at the end of the chapter. The only exceptions to this rule are the extracts themselves, which are referred to by their chapter and letter. The figures in each chapter are similarly labelled in a single sequence. Terminology relating to graph theory, defined in the commentary, is printed in bold type; definitions which occur in the extracts, and those which occur in the commentary but which are not directly relevant to graph theory, are printed in italics. The omission of a section from an abstract is signified by a row of five asterisks; there are also occasional omissions of words or phrases, and these are indicated by five dots. Additions or alterations to the extracts are enclosed in square brackets.

At the end of the book there are three appendices. The first of these gives a brief account of the development of graph theory since 1936. The second appendix contains biographical information about the main characters in our story; the reader should note that capital letters are used in the text for the first mention of all those whose biographies are given in this appendix. The third appendix is a comprehensive bibliography of the work on graph theory published in the period 1736–1936.

A book of this kind could not be written without help, in the form of advice, criticism, and information, from many friends and colleagues. Our special thanks are due to P. J. Federico, who has spent several years working on the history of graph theory. He is preparing a book on the subject which, however, will differ from ours, both in the period covered and in the treatment of the material. Mr. Federico has been generous enough to provide us with drafts of his work, and he has commented in detail on our text. We have pleasure in thanking him for his assistance.

Our thanks are also due to many others who have assisted our work, and especially to D. D. V. Morgan, G. de Barra, R. V. Turley, M. Askew and the staffs of the various libraries who have unearthed a number of obscure journals to meet our requests. We also thank the secretarial staffs of the Mathematics Departments of Royal Holloway College and the Open University for their help with typing the various drafts of the manuscript.

Contents

List of extracts

List of plates

(The plates will be found between pages 208 and 209.)

Acknowledgements

We should like to express our thanks to the following for permission to use copyright material:

Extracts: The Oxford University Press (6D, 7A), the American Mathematical Society (8B, 8D, 9D), Springer-Verlag, Heidelberg (7B, 10D), B. G. Teubner, Stuttgart (3D, 7C), Fundamenta Mathematicae, Warsaw (8C), the Princeton University Press (9A, 9B, 10C), the Johns Hopkins University Press (9C).

Photographs: The American Mathematical Society (photograph of G. D. Birkhoff, from 'Collected Works, Vol. 1'), the Bolyai János Mathematical Society (photograph of D. König), the Mansell Collection (photograph of Cayley), Purnell and Sons, Cape Town (photograph of F. Guthrie from 'Ericas in Southern Africa'), G. Heawood (photograph of P. J. Heawood), the British Museum (map of Königsberg).

1

Paths

J. B. Listing(1808–82)

The origins of graph theory are humble, even frivolous. Whereas many branches of mathematics were motivated by fundamental problems of calculation, motion, and measurement, the problems which led to the development of graph theory were often little more than puzzles, designed to test the ingenuity rather than to stimulate the imagination. But despite the apparent triviality of such puzzles, they captured the interest of mathematicians, with the result that graph theory has become a subject rich in theoretical results of a surprising variety and depth.

In this chapter we shall be concerned mainly with the origin and ramifications of one particular puzzle—the problem of the Königsberg. bridges. The solution of this problem involves the formulation of several of the basic concepts of graph theory.

The problem of the Königsberg bridges

The map in Fig. 1.1 is taken from a book [1] published in the seventeenth century. It is an artist's impression of the old city of Königsberg in Eastern Prussia, showing the River Pregel which flows through the city. (Königsberg is now known as Kaliningrad, and the Pregel is called the Pregolya.) As can be seen, the Pregel surrounds an island (called Kneiphof), and, on the right of the map, it separates into two branches. To enable the citizens of Königsberg to travel easily from one part of the city to another, the river was spanned by seven bridges, with such delightful names as Honey Bridge and Blacksmith's Bridge.

It is said that the people of Königsberg used to entertain themselves by trying to devise a route around the city which would cross each of the seven bridges just once. Since their attempts had always failed, many of them believed that the task was impossible, but it was not until 1736 that the problem was treated from a mathematical point of view and the

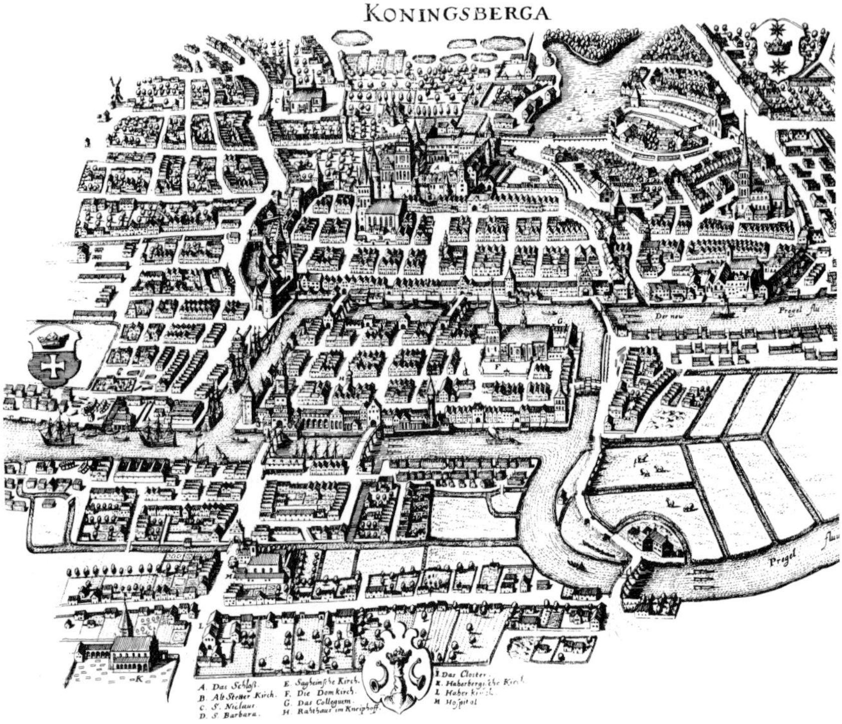

KONINGSBERGA

FIG. 1.1.

impossibility of finding such a route was proved. In that year, one of the leading mathematicians of the time, Leonhard EULER, wrote an article in which he not only dealt with this particular problem, but also gave a general method for other problems of the same type. His article was of considerable importance, both for graph theory and for the development of mathematics as a whole, and we shall give a translation of it in full [1A]. However, the reader may prefer to stop after Paragraph 9, and go on to our commentary at the end of the article, since Euler's main results will be proved more succinctly later in the chapter.

1A

L. EULER

SOLUTIO PROBLEMATIS AD GEOMETRIAM SITUS PERTINENTIS

[The solution of a problem relating to the geometry of position]

Commentarii Academiae Scientiarum Imperialis Petropolitanae **8** (1736), 128–140.

1. In addition to that branch of geometry which is concerned with magnitudes, and which has always received the greatest attention, there is another branch, previously almost unknown, which Leibniz first mentioned, calling it the *geometry of position*. This branch is concerned only with the determination of position and its properties; it does not involve measurements, nor calculations made with them. It has not yet been satisfactorily determined what kind of problems are relevant to this geometry of position, or what methods should be used in solving them. Hence, when a problem was recently mentioned, which seemed geometrical but was so constructed that it did not require the measurement of distances, nor did calculation help at all, I had no doubt that it was concerned with the geometry of position—especially as its solution involved only position, and no calculation was of any use. I have therefore decided to give here the method which I have found for solving this kind of problem, as an example of the geometry of position.

2. The problem, which I am told is widely known, is as follows: in Königsberg in Prussia, there is an island *A*, called *the Kneiphof;* the river which surrounds it is divided into two branches, as can be seen in Fig. [1.2], and these branches are crossed by seven bridges, *a, b, c, d, e, f* and *g*. Concerning these bridges, it was asked whether anyone could arrange a route in such a way that he would cross each bridge once and only once. I was told that some people asserted that this was impossible, while others were in doubt; but nobody would actually assert that it could be done. From this, I have formulated the general problem: whatever be the arrangement and division of the river into branches, and however many bridges there be, can one find out whether or not it is possible to cross each bridge exactly once?

3. As far as the problem of the seven bridges of Königsberg is concerned, it can be solved by making an exhaustive list of all possible routes, and then finding

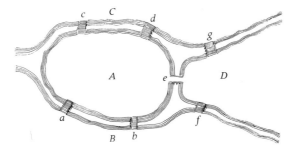

FIG. 1.2.

whether or not any route satisfies the conditions of the problem. Because of the number of possibilities, this method of solution would be too difficult and laborious, and in other problems with more bridges it would be impossible. Moreover, if this method is followed to its conclusion, many irrelevant routes will be found, which is the reason for the difficulty of this method. Hence I rejected it, and looked for another method concerned only with the problem of whether or not the specified route could be found; I considered that such a method would be much simpler.

4.   My whole method relies on the particularly convenient way in which the crossing of a bridge can be represented. For this I use the capital letters A, B, C, D, for each of the land areas separated by the river. If a traveller goes from A to B over bridge a or b, I write this as AB—where the first letter refers to the area the traveller is leaving, and the second refers to the area he arrives at after crossing the bridge. Thus if the traveller leaves B and crosses into D over bridge f, this crossing is represented by BD, and the two crossings AB and BD combined I shall denote by the three letters ABD, where the middle letter B refers both to the area which is entered in the first crossing and to the one which is left in the second crossing.

5.   Similarly, if the traveller goes on from D to C over the bridge g, I shall represent these three successive crossings by the four letters $ABDC$, which should be taken to mean that the traveller, starting in A, crosses to B, goes on to D, and finally arrives in C. Since each land area is separated from every other by a branch of the river, the traveller must have crossed three bridges. Similarly, the successive crossing of four bridges would be represented by five letters, and in general, however many bridges the traveller crosses, his journey is denoted by a number of letters one greater than the number of bridges. Thus the crossing of seven bridges requires eight letters to represent it.

6.   In this method of representation, I take no account of the bridges by which the crossing is made, but if the crossing from one area to another can be made by several bridges, then any bridge can be used, so long as the required area is reached. It follows that if a journey across the seven bridges [of Fig. 1.2] can be arranged in such a way that each bridge is crossed once, but none twice, then the route can be represented by eight letters which are arranged so that the letters A and B are next to each other twice, since there are two bridges, a and b, connecting the areas A and B; similarly, A and C must be adjacent twice in the series of eight letters, and the pairs A and D, B and D, and C and D must occur together once each.

7.   The problem is therefore reduced to finding a sequence of eight letters, formed from the four letters A, B, C and D, in which the various pairs of letters occur the required number of times. Before I turn to the problem of finding such a sequence, it would be useful to find out whether or not it is even possible to arrange the letters in this way, for if it were possible to show that there is no such arrangement, then any work directed towards finding it would be wasted. I have therefore tried to find a rule which will be useful in this case, and in others, for determining whether or not such an arrangement can exist.

8.   In order to try to find such a rule, I consider a single area A, into which there lead any number of bridges a, b, c, d, etc. (Fig. [1.3]). Let us take first the single bridge a which leads into A: if a traveller crosses this bridge, he must either have

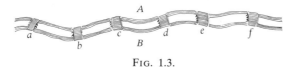

FIG. 1.3.

been in A before crossing, or have come into A after crossing, so that in either case the letter A will occur once in the representation described above. If three bridges (a, b and c, say) lead to A, and if the traveller crosses all three, then in the representation of his journey the letter A will occur twice, whether he starts his journey from A or not. Similarly, if five bridges lead to A, the representation of a journey across all of them would have three occurrences of the letter A. And in general, if the number of bridges is any odd number, and if it is increased by one, then the number of occurrences of A is half of the result.

9.   In the case of the Königsberg bridges, therefore, there must be three occurrences of the letter A in the representation of the route, since five bridges (a, b, c, d, e) lead to the area A. Next, since three bridges lead to B, the letter B must occur twice; similarly, D must occur twice, and C also. So in a series of eight letters, representing the crossing of seven bridges, the letter A must occur three times, and the letters B, C and D twice each—but this cannot happen in a sequence of eight letters. It follows that such a journey cannot be undertaken across the seven bridges of Königsberg.

10.   It is similarly possible to tell whether a journey can be made crossing each bridge once, for any arrangement of bridges, whenever the number of bridges leading to each area is odd. For if the sum of the number of times each letter must occur is one more than the number of bridges, then the journey can be made; if, however, as happened in our example, the number of occurrences is greater than one more than the number of bridges, then such a journey can never be accomplished. The rule which I gave for finding the number of occurrences of the letter A from the number of bridges leading to the area A holds equally whether all of the bridges come from another area B, as shown in Fig. [1.3], or whether they come from different areas, since I was considering the area A alone, and trying to find out how many times the letter A must occur.

11.   If, however, the number of bridges leading to A is even, then in describing the journey one must consider whether or not the traveller starts his journey from A; for if two bridges lead to A, and the traveller starts from A, then the letter A must occur twice, once to represent his leaving A by one bridge, and once to represent his returning to A by the other. If, however, the traveller starts his journey from another area, then the letter A will only occur once; for this one occurrence will represent both his arrival in A and his departure from there, according to my method of representation.

12.   If there are four bridges leading to A, and if the traveller starts from A, then in the representation of the whole journey, the letter A must occur three times if he is to cross each bridge once; if he begins his walk in another area, then the letter A will occur twice. If there are six bridges leading to A, then the letter A will occur four times if the journey starts from A, and if the traveller does not start by leaving A, then it must occur three times. So, in general, if the number of bridges is even, then the number of occurrences of A will be half of this number if

the journey is not started from *A*, and the number of occurrences will be one greater than half the number of bridges if the journey does start at *A*.

13.   Since one can start from only one area in any journey, I shall define, corresponding to the number of bridges leading to each area, the number of occurrences of the letter denoting that area to be half the number of bridges plus one, if the number of bridges is odd, and if the number of bridges is even, to be half of it. Then, if the total of all the occurrences is equal to the number of bridges plus one, the required journey will be possible, and will have to start from an area with an odd number of bridges leading to it. If, however, the total number of letters is one less than the number of bridges plus one, then the journey is possible starting from an area with an even number of bridges leading to it, since the number of letters will therefore be increased by one.

14.   So, whatever arrangement of water and bridges is given, the following method will determine whether or not it is possible to cross each of the bridges:

I first denote by the letters *A*, *B*, *C*, etc. the various areas which are separated from one another by the water. I then take the total number of bridges, add one, and write the result above the working which follows. Thirdly, I write the letters *A*, *B*, *C*, etc. in a column, and write next to each one the number of bridges leading to it. Fourthly, I indicate with an asterisk those letters which have an even number next to them. Fifthly, next to each even one I write half the number, and next to each odd one I write half the number increased by one. Sixthly, I add together these last numbers, and if this sum is one less than, or equal to, the number written above, which is the number of bridges plus one, I conclude that the required journey is possible. It must be remembered that if the sum is one less than the number written above, then the journey must begin from one of the areas marked with an asterisk, and it must begin from an unmarked one if the sum is equal.

Thus in the Königsberg problem, I set out the working as follows:

Number of bridges 7, which gives 8

Bridges

A,	5		3
B,	3		2
C,	3		2
D,	3		2

Since this gives more than 8, such a journey can never be made.

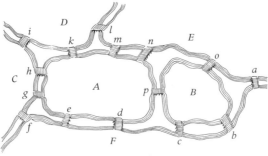

Fig. 1.4.

15. Suppose that there are two islands A and B surrounded by water which leads to four rivers as shown in Fig. [1.4]. Fifteen bridges (a, b, c, d, etc.) cross the rivers and the water surrounding the islands, and it is required to determine whether one can arrange a journey which crosses each bridge exactly once. First, therefore, I name all the areas separated by water as A, B, C, D, E, F, so that there are six of them. Next, I increase the number of bridges (15) by one, and write the result (16) above the working which follows.

$$
\begin{array}{ccc}
 & & 16 \\ \hline
A^*, & 8 & 4 \\
B^*, & 4 & 2 \\
C^*, & 4 & 2 \\
D, & 3 & 2 \\
E, & 5 & 3 \\
F^*, & 6 & 3 \\ \hline
 & & 16
\end{array}
$$

Thirdly, I write the letters A, B, C, etc. in a column, and write next to each one the number of bridges which lead to the corresponding area, so that eight bridges lead to A, four to B, and so on. Fourthly, I indicate with an asterisk those letters which have an even number next to them. Fifthly, I write in the third column half the even numbers in the second column, and then I add one to the odd numbers and write down half the result in each case. Sixthly, I add up all the numbers in the third column in turn, and I get the sum 16; since this is equal to the number (16) written above, it follows that the required journey can be made if it starts from area D or E, since these are not marked with an asterisk. The journey can be done as follows:

$$EaFbBcFdAeFfCgAhCiDkAmEnApBoElD,$$

where I have written the bridges which are crossed between the corresponding capital letters.

16. In this way it will be easy, even in the most complicated cases, to determine whether or not a journey can be made crossing each bridge once and once only. I shall, however, describe a much simpler method for determining this which is not difficult to derive from the present method, after I have first made a few preliminary observations. First, I observe that the numbers of bridges written next to the letters A, B, C, etc. together add up to twice the total number of bridges. The reason for this is that, in the calculation where every bridge leading to a given area is counted, each bridge is counted twice, once for each of the two areas which it joins.

17. It follows that the total of the numbers of bridges leading to each area must be an even number, since half of it is equal to the number of bridges. This is impossible if only one of these numbers is odd, or if three are odd, or five, and so on. Hence if some of the numbers of bridges attached to the letters A, B, C, etc. are odd, then there must be an even number of these. Thus, in the Königsberg problem, there were odd numbers attached to the letters A, B, C and D, as can be seen from Paragraph 14, and in the last example (in Paragraph 15), only two numbers were odd, namely those attached to D and E.

18.   Since the total of the numbers attached to the letters *A, B, C,* etc. is equal to twice the number of bridges, it is clear that if this sum is increased by 2 and then divided by 2, then it will give the number which is written above the working. If, therefore, all of the numbers attached to the letters *A, B, C, D,* etc. are even, and half of each of them is taken to obtain the numbers in the third column, then the sum of these numbers will be one less than the number written above. Whatever area marks the beginning of the journey, it will have an even number of bridges leading to it, as required. This will happen in the Königsberg problem if the traveller crosses each bridge twice, since each bridge can be treated as if it were split in two, and the number of bridges leading into each area will therefore be even.

19.   Furthermore, if only two of the numbers attached to the letters *A, B, C,* etc. are odd, and the rest are even, then the journey specified will always be possible if the journey starts from an area with an odd number of bridges leading to it. For, if the even numbers are halved, and the odd ones are increased by one, as required, the sum of their halves will be one greater than the number of bridges, and hence equal to the number written above.

It can further be seen from this that if four, or six, or eight ... odd numbers appear in the second column, then the sum of the numbers in the third column will be greater by one, two, three ... than the number written above, and the journey will be impossible.

20.   So whatever arrangement may be proposed, one can easily determine whether or not a journey can be made, crossing each bridge once, by the following rules:

If there are more than two areas to which an odd number of bridges lead, then such a journey is impossible.

If, however, the number of bridges is odd for exactly two areas, then the journey is possible if it starts in either of these areas.

If, finally, there are no areas to which an odd number of bridges leads, then the required journey can be accomplished starting from any area.

With these rules, the given problem can always be solved.

21.   When it has been determined that such a journey can be made, one still has to find how it should be arranged. For this I use the following rule: let those pairs of bridges which lead from one area to another be mentally removed, thereby considerably reducing the number of bridges; it is then an easy task to construct the required route across the remaining bridges, and the bridges which have been removed will not significantly alter the route found, as will become clear after a little thought. I do not therefore think it worthwhile to give any further details concerning the finding of the routes.

Euler's treatment of the Königsberg problem involved two major steps. First, in Fig. 1.2, he replaced the map of the city by a simple diagram showing its main features, and then, in Paragraphs 4 and 7 of his article, he formulated the problem in such a way that the diagram became

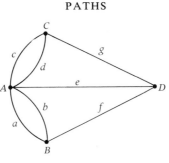

F<small>IG</small>. 1.5.

unnecessary. He denoted the four land areas by the symbols A, B, C, D, and the seven bridges by a, b, c, d, e, f, g, where the bridge a joins A and B, e joins A and D, and so on. This is an example of what we now refer to as a 'graph', and Euler's problem of finding a sequence of eight symbols with a particular property (described in Paragraph 7) is related to the existence of a special kind of 'path' in the graph.

To explain the exact meaning of these terms we must give some definitions. A **graph** consists of a finite set of **vertices**, a finite set of **edges**, and a rule which tells us which edges **join** which pairs of vertices. Normally, an edge joins two distinct vertices, but exceptionally the two vertices may coincide; in the latter case the edge is said to be a **loop**. In our particular example there are four vertices, corresponding to the land areas A, B, C, D, and seven edges, corresponding to the seven bridges; the rule tells us that the edges a and b join the vertices A and B, the edges c and d join the vertices A and C, and so on. We also define a **path** in a graph to be a sequence of vertices and edges,

$$v_0, e_1, v_1, e_2, v_2, \ldots, v_{r-1}, e_r, v_r,$$

in which each edge e_i joins the vertices v_{i-1} and v_i $(1 \le i \le r)$.

It is helpful to illustrate these abstract definitions by representing a graph pictorially. We depict a graph as a diagram of points and lines, in which the points represent vertices and the lines represent edges; a diagram for the Königsberg graph is shown in Fig. 1.5. It should be noted that this is merely a convenient way of describing the graph—we repeat that the graph itself is an abstract entity consisting of the four vertices A, B, C, D, the seven edges a, b, c, d, e, f, g, and the rule which tells us how the edges join the vertices. Nevertheless, the pictorial representation of graphs is a very useful technique and we shall use it throughout this book.

We may now formulate the problem of the Königsberg bridges using the terminology just introduced: the object of the problem is to find a path which contains each edge of the graph once and only once. A path of this kind is now called an **Eulerian path**, and Euler showed that the

Königsberg graph has no such path. He also investigated the existence of Eulerian paths in general graphs.

In order that a graph should contain an Eulerian path, it is clearly necessary that the graph should be **connected**; this means that for any two vertices v and w it is possible to find a path beginning at v and ending at w, so that the graph is 'all in one piece'. Euler took this condition for granted, since it was automatically satisfied in the examples he considered. A **disconnected** graph, that is, one which is not connected, splits up into connected parts, called its **components**.

We need only one more definition at this stage. The **valency** (or **degree**) of a vertex v is the number of edges which meet at v; for example, in the Königsberg graph, the valency of the vertex A is five, and the valencies of B, C, and D are all three. (The reason for the name 'valency' will be given in Chapter 4.) Notice that if we add the valencies of all the vertices in a graph, then the sum is just twice the number of edges in the graph, since each edge contributes twice to the sum. This fact was mentioned in Paragraph 16 of Euler's article, and it yields the useful result (Paragraph 17) that, *in any graph, the number of vertices with odd valency must be even.* This is sometimes referred to as the 'handshaking lemma', since it tells us that if the guests at a party shake hands when they meet, then the number of guests who have shaken hands an odd number of times must necessarily be even.

We now have all the terminology we need to state Euler's main result, given in lines 4 and 5 of Paragraph 20: *if a connected graph has more than two vertices of odd valency, then it cannot contain an Eulerian path.* In the same paragraph, Euler also stated the converse result: *if a connected graph has no vertices of odd valency, or two such vertices, then it contains an Eulerian path.* Unfortunately, Euler did not give a proof of the latter result, presumably because he considered it to be self-evident. This lack of precision was quite common among eighteenth-century mathematicians, and occasionally it led them into the realm of fantasy, as when Euler asserted the truth of the equation

$$1 - 1 + 1 - 1 + 1 - \ldots = \tfrac{1}{2}.$$

Nevertheless, Euler's graph-theoretical intuition was correct, although a complete proof of the converse result did not appear in print until 1873 [1B]. The proof was due to a young German mathematician, Carl HIERHOLZER, whose work was prepared for publication by a colleague, C. Wiener. The tragic circumstances were explained in a footnote:

> Privatdocent Dr. Hierholzer, unfortunately prematurely snatched away by death from the service of scholarship (died 13 September 1871), reported on the following investigation to a circle of mathematical friends. It was in

order to save it from oblivion (and it had to be reconstructed without any written record) that I sought to complete the following as accurately as possible, with the help of my esteemed colleague Lüroth.

1B

C. HIERHOLZER
ÜBER DIE MÖGLICHKEIT, EINEN LINIENZUG OHNE
WIEDERHOLUNG UND OHNE UNTERBRECHNUNG ZU UMFAHREN

[On the possibility of traversing a line-system without repetition or discontinuity]

Mathematische Annalen **6** (1873), 30–32.

In an arbitrary system of interwoven lines, we can define the *branches* at a point to be the distinct lines of the network along which it is possible to leave the point in question. A point at which there are several branches is called a *node*, and is termed as many-fold as the number of branches there, being called odd or even according as this number is odd or even. Thus, an ordinary double point may be called a four-fold node, an ordinary point is a two-fold node, and a free end may be termed a one-fold node.

If a line-system can be traversed in one path without any section of line being traversed more than once, then the number of odd nodes is either zero or two. If, in carrying out this process, we pass through any node, then two of the branches at that node are used, and since no line-segment may be traversed twice, a node which we pass through n times must be a $2n$-fold node. A point can therefore be an odd node only if on one occasion we do not pass through it, that is, if it is an initial or terminal point. If, on reaching the end of the journey, we return to the starting point, then there can only be even nodes; if not, then the initial and terminal points are odd nodes.

Conversely: *if a connected line-system has either no odd nodes or two odd nodes, then the system can be traversed in one path.*

For (a) if only part of the line-system has been traversed, then every node in the remaining part remains even or odd, just as it was in the original system; only the initial and terminal nodes of the traversed part change their parity, unless they coincide. This is because two branches are used in passing through a node, and only one branch is used at the start and finish of the path.

(b) If we begin to traverse the system at an odd node, then we can finish only at another odd node. This is because two branches are used each time we pass through an even node, so that each time we arrive at such a node, there is at least one other branch available to depart along. However, the initial node is converted at the beginning to an even node, so that it is also impossible to stop there. On the other hand, if we start to traverse the system at an even node, then we can also terminate at the same node, since it is changed at the outset to an odd one.

(c) If, now, the system has two odd nodes, then a path beginning at one of them necessarily terminates at the other. In this case the completed part of the path is open. If, on the other hand, the given system has no odd nodes, then a path beginning at any node (which must be an even node) must necessarily terminate at that same node. In this case the completed path is closed.

(d) If a part *b* remains untraversed, then it can contain only even nodes, since the two odd nodes possibly available are used in the first part, and the remaining nodes retain their parity. Similarly, *b* must meet the part *a* already traversed in at least one common node, since otherwise the system would break up into several separate pieces. If we proceed from the part *a* at such a node *P* into the part *b*, we must necessarily terminate in *b* at *P*, and so we can continue from there to traverse *a*, as we showed earlier. In this way, each part not yet traversed can be attached to the part already traversed, and the whole system can be covered in one path.

This also gives us the following theorem: *A line-system contains an even number of odd nodes.* If we eliminate a path which begins at an odd node and continues as far as possible until it can go no further (which again must happen at an odd node), then two odd nodes are thereby removed. By repeating this process, we can reduce the number of odd nodes to fewer than two. Now this remainder cannot be one, but must be zero; this is because if a system contained just one odd node and a path were to begin from it, then it would never end, since this can happen only at an odd node. The number of nodes of the original system is therefore even.

Editors' note. The essential content of the above, but in abbreviated form, and in places with incomplete proof, is to be found in the unfortunately little-known treatise of Listing, *Vorstudien zur Topologie.* Possibly the above exposition may serve to guide the attention of geometers towards this work, important also in several other connections.

Hierholzer himself was apparently unaware of Euler's work, and the editorial note refers only to a treatise published in 1847 by Johann Benedict LISTING, of which we shall have more to say shortly. Such ignorance of earlier work is not surprising when one considers the difficulties of all forms of communication in those times; clearly, much of mathematical lore was transmitted orally or by private correspondence, resulting in a certain lack of reliability. Fortunately, Euler's work had not been completely forgotten, for in 1851 a French translation [2] by É. Coupy was published in the *Nouvelles Annales de Mathématiques*, a journal intended primarily for the students at the École Polytechnique in Paris. Coupy also applied the methods of Euler to the analogous problem of the bridges over the River Seine. Nor was the problem forgotten in Königsberg; in 1875, L. Saalschütz [3] reported that a new bridge had been constructed there, joining the land areas denoted by *B* and *C*, and that the citizens' perambulation was now theoretically possible.

Diagram-tracing puzzles

Coupy's interest in the work of Euler stemmed from two references to it in the early nineteenth century. The first of these occurred in a textbook of algebra [4], written by the Swiss mathematician Simon-Antoine-Jean LHUILIER in 1804, and the second appeared in a memoir on

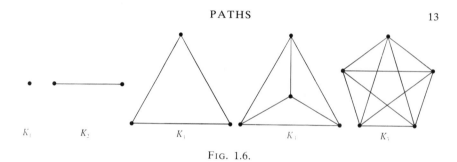

FIG. 1.6.

polygons and polyhedra [5], written by Louis POINSOT in 1809. The latter work dealt with a variety of geometrical problems, one of which was the following:

> Given some points situated at random in space, it is required to arrange a single flexible thread uniting them two by two in all possible ways, so that finally the two ends of the thread join up, and so that the total length is equal to the sum of all the mutual distances. As we shall see, the solution is possible only for an odd number of points . . .

The problem, expressed in modern terminology, concerns the existence of an Eulerian path in a special class of graphs, known as the complete graphs. For each natural number n, the **complete graph** K_n on n vertices is constructed by joining each pair of vertices by an edge, so that there are $\frac{1}{2}n(n-1)$ edges in all (see Fig. 1.6). Since each vertex is joined to all the others, the graph K_n has n vertices of valency $n-1$. Poinsot proved, using arguments similar to those of Euler, that an Eulerian path in K_n is impossible when $n = 4, 6, 8, \ldots$, because in these cases there are more than two vertices with odd valency. He also gave an ingenious method for constructing an Eulerian path in K_n, when n is odd.

He went on to remark that such methods could be applied to

> . . . a little positional puzzle which was spoken of a few years ago. It is proposed to describe the four sides and the two diagonals of a quadrilateral in a simple continuous stroke. It may be seen that this is impossible; for as there are three adjacent lines at each angle, one of them is necessarily the continuation of another; but as for the third, its extremity must be a point at which the pen will stop or start: thus, there must be at least four points where the pen is stationary; and so the figure cannot be drawn with fewer than two separate strokes.

Problems of this kind may often be found in books of mathematical puzzles. We are given a diagram, such as that shown in Fig. 1.7, and we are asked to draw it using as few continuous pen-strokes as possible, and without going over any part of it twice. For example, it is easily seen that the diagram in Fig. 1.7 requires only two separate strokes. Clearly, a diagram can be drawn with one stroke if and only if the corresponding graph has an Eulerian path. Poinsot explained how such problems may be

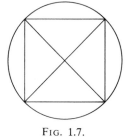

F<small>IG</small>. 1.7.

solved, and he gave some examples—for instance, he pointed out that four separate strokes are required to cover the edges of a cube.

Another treatment of this theory appeared in 1847 in the important treatise by Listing, mentioned in the editorial footnote to Hierholzer's paper. Listing set out to investigate those aspects of geometry which depend upon inherent positional relationships, rather than the measurement of lengths and angles. He discussed such things as screws, knots, and links, and he devoted a short section [1C] of his work to diagram-tracing puzzles. It must be said that his account adds very little to those previously given by Euler and Poinsot, but, in view of the historical importance of his treatise, it merits the inclusion of an extract in this book.

1C

J. B. LISTING

VORSTUDIEN ZUR TOPOLOGIE

[Introductory studies in topology]

Göttinger Studien **1** (1847), 811–875.

*   *   *   *   *

Linear complexes—that is, any straight lines, curved lines, or any aggregate of such lines—can either be contained in a surface (a plane or a spherical surface) or else may twist through space in every direction. In both cases there can be an arbitrary number of meeting-points or crossings, just as in the second case, if we project onto a plane or spherical surface, there will also arise an arbitrary number of 'crossing-over' points in addition to the intersections.

In the case of linear complexes in a surface, where consequently only meeting-points or crossings occur, each point belonging to the linear complex can be regarded as a place from which there are available a certain number of routes to other points of the complex. For example, on a straight line drawn between two points *A* and *B*, there is just one route available at each of the endpoints *A* and *B*, whereas at each internal point *C* of the line there are two routes. If we now suppose that no route is covered more than once, then as before A and B must be the endpoints of a continuous simple path, whereas we can arrive at the point C from one direction and leave again in the other direction.

It is clear that if n is the number of routes at an intersection point, then the point may be considered either as the crossing-point of $\frac{1}{2}n$ lines or as the crossing-point of $\frac{1}{2}(n-1)$ lines together with the endpoint of one line, according as n is even or odd. From this it follows that a complete linear complex can be regarded as the aggregate of a certain number of continuous strokes, each of which has two boundary (or end) points which lie on an intersection of odd-numbered type. Thus, in every linear complex (and, likewise, in every union of mutually disjoint complexes), there must occur an even number of intersections of odd type, whereas there must be infinitely many points of even type, since each internal point of any line is such a point. With the help of this result, it is easy to find, in any given complex, the minimum number of continuous strokes in which such a complex, however complicated, can be drawn without any part of it being covered more than once. In order to determine this minimum number of strokes, we have accordingly to count the odd points of the complex. Let this number, which must necessarily be even, be p; then $\frac{1}{2}p$ is the least number of strokes in which the configuration can be drawn. The actual construction, as well as the determination of the number of possible solutions, would certainly entail more complicated investigations than there is room for here.

It remains to be mentioned that if $p = 0$, the number of strokes is to be set equal to one, rather than zero, and that these requirements may be applied separately to each individual one of any given number of mutually disjoint complexes. Thus, for example, in order to draw m mutually disjoint complexes in which there are no points of odd type (as, for example, when the m parts lie concentrically in one another like separate circuits), m strokes are required. The same is true for drawing a complex with $2m$ intersections of odd type, or m mutually disjoint complexes each of which contains two odd points; and further, that the success of the actual construction when $p \neq 0$ requires the condition that the beginning and end of each stroke occur only at odd intersections.

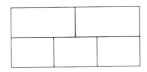

FIG. 1.8.

In Fig. [1.8] there are eight odd points (each on three routes), so that this figure cannot be drawn with fewer than four continuous strokes without covering some part more than once [6].

A square with a circumscribed circle and both diagonals included requires at least two strokes, and the lattice of a chessboard with 64 squares requires at least 14. If each of the black squares of the chessboard is given one diagonal, all sloping in the same direction, then the construction can be carried out in either 9 or 6 continuous strokes, depending on whether or not the diagonals of the black corner squares lie in a direct line. If all of the squares are given such mutually parallel diagonals, then one can complete the drawing of the figure in one single stroke.

FIG. 1.9.

Fig. [1.9] can likewise be drawn in a single stroke, since it has only two points of odd type, both five-fold, lying at the ends of the horizontal axis.

*   *   *   *   *

In 1849 a commentary on Poinsot's memoir appeared in the *Nouvelles Annales de Mathématiques* [7]. Among other things, it drew attention to an interesting interpretation of Poinsot's rule for an Eulerian path in K_7, in terms of the game of dominoes. If we set aside the doubles (that is, the dominoes 0–0, 1–1, ..., 6–6), the remaining twenty–one dominoes correspond to the edges of K_7, and an Eulerian path represents a sequence of the twenty-one dominoes with all of the ends matching up, as required by the rules. Since the doubles can be inserted at appropriate places once the basic sequence is given, we have proved that a complete game of dominoes is possible.

The question of finding the number of ways in which this can happen became known as the 'domino problem'; it aroused a certain amount of interest, but no solution appeared until 1871 [8]. A general method for counting the number of Eulerian paths in a graph was later given by Gaston TARRY in 1886 [9].

The simple theory of Euler, Poinsot, Listing, and Hierholzer has been rediscovered many times. As late as 1905, John Cook Wilson, Wykeham Professor of Logic at Oxford University, published an incredibly long-winded treatise [10] on the subject; he made no reference whatever to previous related work, of which he was apparently quite unaware.

Mazes and labyrinths

There is yet another problem which is related to those we have been considering: it is the question of how to escape from a maze or labyrinth. This problem and its solution are now quite well known, since they are discussed in many books of mathematical puzzles, in particular, the three 'classics': É. Lucas' *Récréations mathématiques* [11], W. W. Rouse Ball's *Mathematical recreations and problems* [12] (later entitled *Mathematical*

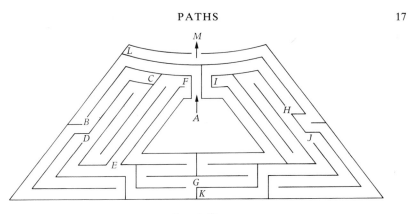

Fɪɢ. 1.10.

recreations and essays), and W. Ahrens' *Mathematische Unterhaltungen und Spiele* [13]. These books contain historical material which indicates that the construction of mazes is an ancient art, but it seems that no careful mathematical account of their properties was published before the latter part of the nineteenth century.

In order to explain the mathematical content of the maze problem, let us consider the famous maze at Hampton Court in England (Fig. 1.10). The essential features of this maze are a number of passages or pathways, separated by hedges, and meeting at a certain number of junctions. We may represent any such maze by a graph in which the vertices correspond to the junctions and the edges correspond to the passages, so that the graph corresponding to the Hampton Court maze is as depicted in Fig. 1.11.

In graph-theoretical terms the maze problem requires a path, which starts from some given vertex (the centre), and which ends at another given vertex (the exit). In fact, it is sufficient to be less specific and merely to ask for a path which contains every edge of the graph, since then both the centre and the exit will be visited at some stage. This latter problem always has a solution, provided that the graph is connected. To see this, we replace every edge in the original graph by two edges joining the same pair of vertices; the result is a new graph in which every vertex has even valency, and which, consequently, must have an Eulerian path. In the

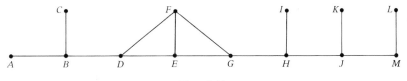

Fɪɢ. 1.11.

original graph we obtain a path which contains each edge exactly twice, and this is sufficient for our purposes.

Unfortunately, the preceding proof is merely a proof of the existence of a path, and as such it affords but little solace to any one who is lost in a maze without a plan of its layout. What one requires in those circumstances is a practical means of escaping—that is, a constructive proof of the existence of the desired path. Mathematicians call such a plan of campaign an 'algorithm'; in this case the algorithm must take the form of a rule which can be applied on arrival at each junction. We emphasize that a map of the maze is not provided, but it is presumed that the route can be marked as it is traversed, if this is desirable.

The first attempt to give an algorithm for the maze problem was published in 1873 [14] by Wiener, in the same issue of *Mathematische Annalen* as Hierholzer's posthumous paper. Wiener's paper was unnecessarily complicated—for instance, he devoted a paragraph to proving that after entering a maze one can always return to the entrance; as KÖNIG later pointed out [15, p. 36], this limited objective can be achieved by a somewhat simpler rule: don't go in! Wiener's rule for escaping from the centre is also rather involved, and it may entail traversing a passage many times, instead of just twice.

A more efficient algorithm was discovered by Trémaux and published in the first volume of Lucas' book [11] in 1882; in this solution no passage need be traversed more than twice. Regrettably, the proof given by Lucas is seriously deficient, although the algorithm is correct. The simplest, and best, algorithm was published by Tarry in 1895, and his paper [1D] is the last extract in this chapter.

1D

G. TARRY
LE PROBLÈME DES LABYRINTHES
[The maze problem]

Nouvelles Annales de Mathématiques (3) **14** (1895), 187–190.

Any maze can be traversed in a single journey, passing twice (in opposite directions) along each passage, without the necessity of knowing its plan.

To resolve this problem, it is enough to follow this single rule:

Do not return along the passage which has led to a junction for the first time unless you cannot do otherwise.

We begin with some comments.

At any instant, before arriving at a junction or after having left it, every junction other than the point of departure must have an even number of passages which have been traversed just once; and the number of these passages traversed in the inward direction is equal to the number traversed in the outward direction. This is clear, since the number of arrivals is equal to the number of departures, and a passage can be traversed at most twice and in opposite directions. It follows that at the instant of arrival at a junction, the number of passages traversed once only, and in the inward direction, exceeds by one the number of passages traversed once only in the outward direction. If there is only one passage which has been used just once, then this must be the first passage, and all the other passages must necessarily have remained unexplored, or have been traversed twice in opposite directions. Thus one cannot be halted at this junction and one is forced to return along the incoming passage only in the case when all other passages have been traversed twice.

Thus, in following the rule, one can be stopped only at the point of departure, and one returns along the incoming passage at a junction only after all the other passages at that junction have been traversed twice.

Consider now the initial junction. At the instant of an arrival at this junction, the number of passages traversed once in the inward direction is equal to the number of passages traversed once in the outward direction. Consequently, one can be halted only if there are no unexplored passages, and no passages traversed once only. Thus when one is forced to stop at the initial junction, all the passages at that junction have been traversed twice.

Given this, let A, B, C, D, . . . , Z be the initial junction and the other junctions, in the order in which they are visited on the route. If one has followed the single rule strictly, all the passages meeting at the junctions B, C, D, . . . , Z must have been traversed twice, in opposite directions, as must the passages at the initial junction A.

For, since the initial passage AB leading to the junction B from A has been traversed twice in opposite directions, all the passages at the junction B have been traversed twice in opposite directions. Similarly, all the passages at the junction C have been traversed twice, since the incoming passage BC (or AC, if AB is an impasse), which meets the junction B (or A), has been traversed twice. Finally, any junction T, whose incoming passage must necessarily lead from a previous junction (A, B, . . . , or S), has been completely explored, since its incoming passage has been traversed twice. This is what was to be proved.

Suppose that on traversing a passage for the first time one leaves at the entrance two markers, and at the exit one or three markers, depending upon whether the passage leads to a junction already visited or to a new junction; and suppose that on entering a passage where one finds one marker, and so traversing it a second time in the opposite direction, one adds a second marker at the entrance. Then, on arrival at a junction, one can always pick out the new passages (which have no markers), the initial passage (which has three markers), and the other passages traversed once only in the inward direction (which have one marker). The single rule can be stated as follows:

On arriving at a junction, take at random any passage which has no markers or any passage with one marker; if no such passage exists, take the passage with three markers.

By following this practical method, anyone lost in a maze or catacombs must return to the entrance before exploring all the passages and without traversing the same passages more than twice. This shows that it is always possible to escape from a maze, and that the best form of Ariadne's thread is the thread of reason.

Remark.—The general rule allows one to retrace one's steps, when one arrives by a new route at a junction which has already been visited. If this is made obligatory, then on arrival at a junction by a path previously traversed, one will find only a single passage used once, the initial one, and it will be permissible to take this passage only when there are no unexplored passages. It is clear that in this case it becomes unnecessary to distinguish the beginning of the initial passage by a special marker. This particular solution has been found by Trémaux [11].

The problems discussed in this chapter have two desirable features: they may be formulated within a precise mathematical framework, and they admit neat and compact solutions. The framework will be enlarged gradually throughout this book, in order to formulate a wide range of problems from many different sources; but even using this larger vocabulary, it will not be possible to solve all the problems that arise. For example, in the next chapter we shall introduce a problem concerned with the existence of a certain kind of 'circuit' in a graph. Although this problem can be formulated exactly, it will become clear that it does not admit a simple or concise solution. This is all the more surprising since it is superficially very similar to the Eulerian problems discussed in this chapter.

References

1. ZEILLER, M. *Topographia Prussiae et Pomerelliae*. Frankfurt, c.1650.
2. COUPY, É. Solution d'un problème appartenant a la géométrie de situation, par Euler. *Nouv. Ann. Math.* **10** (1851), 106–119.
3. SAALSCHÜTZ, L. [Untitled]. *Schr. Phys.-Ökon. Ges. Königsberg Prussia* **16** (1876), 23–24.
4. LHUILIER, S.-A.-J. *Élémens raisonnés d'algèbre*. Genève, 1804.
5. POINSOT, L. Sur les polygones et les polyèdres. *J. École Polytech.* **4** (Cah. 10) (1810), 16–48.
6. CLAUSEN, T. [Second postscript to:] De linearum tertii ordinis proprietatibus. *Astron. Nachr.* **21** (1844), col. 209–216.
7. TERQUEM, O. Sur les polygones et les polyèdres étoilés, polygones funiculaires; d'après M. Poinsot. *Nouv. Ann. Math.* **8** (1849), 68–74.
8. REISS, M. Evaluation du nombre de combinaisons desquelles les 28 dés d'un jeu du domino sont susceptibles d'après la règle de ce jeu. *Ann. Mat. Pura Appl.* (2) **5** (1871–3), 63–120.
9. TARRY, G. Nombre de manières distinctes de parcourir en une seule course toutes les allées d'un labyrinthe rentrant, en ne passant qu'une seule fois par chacune des allées. *Compt. Rend. Ass. Franç. Avance. Sci.* **15** Pt. 2 (1886), 49–53.
10. WILSON, J. C. *On the traversing of geometrical figures*. Clarendon Press, Oxford, 1905.
11. LUCAS, É. *Récréations mathématiques*, 4 vols. Gauthier–Villars, Paris, 1882–94.
12. ROUSE BALL, W. W. *Mathematical recreations and problems of past and present times*, (1st edn.). Macmillan, London, 1892.
13. AHRENS, W. *Mathematische Unterhaltungen und Spiele*. Leipzig, 1901.
14. WIENER, C. Über eine Aufgabe aus der Geometria situs. *Math. Ann.* **6** (1873), 29–30.
15. KÖNIG, D. *Theorie der endlichen und unendlichen Graphen*. Akademische Verlagsgesellschaft, Leipzig, 1936. Reprinted: Chelsea, New York, 1950.

2
Circuits

T. P. KIRKMAN (1806–95)

THE philosopher and mathematician G. W. Leibniz played an important, and well-known, part in the discovery of the differential calculus. But it is less widely known that to Leibniz should also go the distinction of suggesting one of the most fertile fields of current mathematical research. In a letter [1, pp. 18–19] to C. Huygens, written in 1679, he declared:

> I am not content with algebra, in that it yields neither the shortest proofs nor the most beautiful constructions of geometry. Consequently, in view of this, I consider that we need yet another kind of analysis, geometric or linear, which deals directly with position, as algebra deals with magnitude.

The study of position, which is closely linked with graph theory, is now known by the name suggested by Listing; it is called 'topology'. The subject was slow to capture the interest of mathematicians. Indeed, in 1833, C. F. Gauss wrote [2, p. 605]:

> Of the geometry of position, which Leibniz initiated and to which only two geometers, Euler and Vandermonde, have given a feeble glance, we know and possess, after a century and a half, very little more than nothing.

Undeterred by Gauss' dismissive tone, we have decided to include in this book articles by both Euler and the French mathematician Alexandre-Theophile VANDERMONDE. There is some confusion as to which of Euler's works Gauss intended to cite in his remarks. It is possible that he was referring to the paper [1A] on the Königsberg bridges, but some writers (notably Listing, who had been a student of Gauss), thought that he meant another article [3] whose subject we shall describe below. The work of Vandermonde, although not so famous as that of Euler, is also of considerable interest, and it is our first extract in this chapter.

The knight's tour

Vandermonde's article [2A] is mainly concerned with the problem known as the 'knight's tour' on a chessboard. As is well known, a knight must move two squares parallel to one side of the board, followed by one square in a perpendicular direction; the problem is to find a sequence of moves so that a knight may visit each square of the board just once and finish on the same square as it began. The question has a long history, accounts of which may be found in the books of Ahrens, Lucas, and Rouse Ball listed in the references for Chapter 1. Both Euler and Vandermonde gave systematic treatments of the problem, and Vandermonde used it to illustrate his ideas on the geometry of position.

2A

A.-T. VANDERMONDE
REMARQUES SUR LES PROBLÈMES DE SITUATION
[Remarks on problems of position]

Histoire de l'Académie des Sciences (Paris) (1771), 566–574.

Whatever the twists and turns of a system of threads in space, one can always obtain an expression for the calculation of its dimensions, but this expression will be of little use in practice. The craftsman who fashions a *braid*, a *net*, or some *knots* will be concerned, not with questions of measurement, but with those of position; what he sees there is the manner in which the threads are interlaced. It would therefore be useful to have a system of calculation more relevant to the worker's mode of operation, a notation which would represent his way of thinking, and which could be used for the reproduction of similar objects for all time.

My aim here is simply to demonstrate the possibility of such a notation, and to demonstrate its use in questions concerning networks of threads. In order to illustrate my ideas, I shall consider a well-known problem which belongs to this category, that of the *knight's tour in chess*, solved by Euler in 1759 [3]. The method of that great geometer presupposes that one has a *chessboard* to hand; I have reduced the problem to simple arithmetic, using numbers which do not represent quantities at all, but regions of space.

*   *   *   *   *

(1.)   I consider space divided into arbitrary finite elements, distinguished by their order; that is to say, (1) I consider a plane divided by parallel lines into a series of strips, and then divided again by another set of parallels which cut the first set; I distinguish the different strips by the designation first, second, third, fourth, etc., in both divisions. I can then describe a given point belonging to any one of the

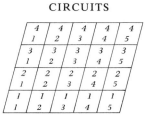

FIG. 2.1.

parallelograms formed by the double divisions, by simply writing two numbers, one above the other, where one number is the order of the first division and the other that of the second. Thus, for example, $\frac{3}{4}$ belongs to the parallelogram which is common to the fourth strip in the first division and the third in the second division (*see figure* [2.1]). (2) I consider a solid space divided first by parallel planes into a series of slices, then divided again by another series of parallel planes which cut the first ones, and finally divided a third time by a new series of parallel planes which cut both of the others. I can then describe by a symbol 1_3^2, for example, a given point belonging to the parallelepiped common to the third slice in the first division, the second slice in the second division, and the first in the third division (*see figure* [2.2]).

(2.) In general, if three numbers arranged as c_a^b designate a certain region of space confined between given limits, then a series of such expressions c_a^b c_a^b, $\gamma_\alpha^\beta, \ldots$ serves to describe a route in space or the path of a thread; and according as one requires greater or lesser precision, one refines or relaxes the limits, and specifies a greater or lesser number of crossings between neighbouring parallelepipeds.

Thus, for example, supposing that the thread is flexible and free to move, except at the points designated below, one has, it seems to me, an expression sufficient for the *plait* (*fig.* [2.3]) by means of the sequence

$$2_2^3, 1_3^2, 2_3^1, 1_1^1, 2_1^2, 1_2^3, \quad 2_4^3, 1_5^2, 2_5^1, 1_3^1, 2_3^2, 1_4^3, \ldots \ldots$$

Similarly, one has an expression for the system of *stocking-stitches* (*fig.* [2.4]) by

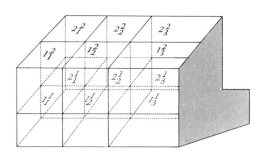

FIG. 2.2.

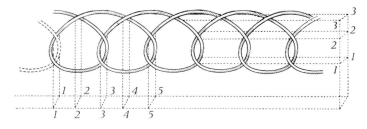

FIG. 2.3.

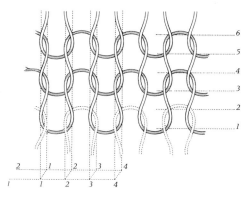

FIG. 2.4.

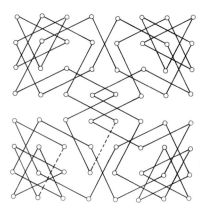

FIG. 2.5.

means of the sequences

$$2_1^4,\ 1_1^3,\ 1_1^2,\ 2_1^1,\ 2_2^1,\ 1_2^2,\ 1_2^3,\ 2_2^4,\ 2_3^4,\ 1_3^3,\ 1_3^2,\ 2_3^1,\ \ldots\ldots$$
$$2_1^6,\ 1_1^5,\ 1_1^4,\ 2_1^3,\ 2_2^3,\ 1_2^4,\ \ldots\ldots$$
$$2_1^8,\ \ldots\ldots$$

&c.

*   *   *   *   *

(3.) The problem of how a *knight* can visit all the squares of a *chessboard*, without visiting the same one twice, reduces to the determination of a certain route of the *knight* on the *chessboard*; or equally, if one supposes that a pin is fixed in the centre of each square, the problem reduces to the determination of a path taken by a thread passed once round each pin and following a rule whose formulation we seek. *See fig.* [2.5].

(4.) If $\frac{b}{a}$ denotes any square of the *board*, then a and b can be any of the numbers 1, 2, 3, 4, 5, 6, 7, 8.

Given two successive positions of a *chess piece*, $\frac{b}{a}$ and $\frac{b}{a}$, the *move* of that *piece* represents a condition on a and a, b and b.

For example, if a *rook* is placed on the square $\frac{b}{a}$, it can go from there to any square $\binom{b}{a+m}$ or $\binom{b+m}{a}$, where m is a positive or negative integer such that neither $a+m$ nor $b+m$ is less than 1 or greater than 8.

Similarly, if a *bishop* is placed on the square $\frac{b}{a}$, it can go to any square $\binom{b\pm m}{a\pm m}$. A *knight* can go from $\frac{b}{a}$ to $\binom{b\pm 1}{a\pm 2}$, or to $\binom{b\pm 2}{a\pm 1}$; and so on for the other pieces.

(5.) The problem of how a *knight* can visit all the squares of the *board*, without visiting the same one twice, thus becomes that of arranging the sixty-four terms

$$\frac{1\ 1\ 1\ 1}{1\ 2\ 3\ 4}\cdots\frac{2\ 2\ 2}{1\ 2\ 3}\cdots\frac{3\ 3}{1\ 2}\cdots\cdots\frac{8}{8}$$

in such a way that the following rule holds for every two adjacent terms: the difference between the upper numbers is 1 and the difference between the lower numbers is 2, or the difference between the upper numbers is 2 and the difference between the lower ones is 1. (*See the example in Section* 11.)

*   *   *   *   *

(7.) To simplify the solution, one can seek a symmetric route for the path of the *knight*; it is from this viewpoint that I shall proceed.

The *knight's* path will be a symmetric figure if, when we interchange the numbers 8 and 1, 7 and 2, 6 and 3, 5 and 4, and *vice versa*, either in the upper sequence, or in the lower, or in both at once, then there is no change in the path as a whole.

(8.) Thus it is required to find sixteen consecutive *knight's* moves, or sixteen terms of the sequence, such that (*see the example below, in Section* 9) if one interchanges 8 and 1, 7 and 2, 6 and 3, 5 and 4, and *vice versa*, in the lower sequence, one gets no term already in the original sixteen, and the same is true if one makes these changes in the upper sequence, or in both the upper and lower sequences simultaneously. After making these transformations, one has four sequences of *knight's* moves, covering the sixty-four squares of the *board* without repetition, and forming a symmetric figure; in order to solve the given problem it remains to join the four sequences into a single one, so that the rule holds at the junctions.

(9.) To obtain the desired sixteen terms, I begin by writing down all sixty-four terms

$$\frac{1\ 1\ 1\ 1}{1\ 2\ 3\ 4} \cdots \frac{2\ 2\ 2}{1\ 2\ 3} \cdots \frac{3\ 3}{1\ 2} \cdots \cdots \frac{8}{8}$$

in any order. I then take one at random, say $\frac{5}{5}$, and write underneath the three transforms $\frac{4}{5}$, $\frac{5}{4}$, and $\frac{4}{4}$. Since these are of no further interest, I delete them from the sixty-four. From the remaining sixty I take at random one which is related to $\frac{5}{5}$ by a *knight's move*, for example $\frac{4}{3}$ (since the difference between the upper numbers is 1, whereas the lower difference is 2). I write underneath $\frac{4}{3}$ the three corresponding transforms $\frac{5}{3}$, $\frac{4}{6}$, $\frac{5}{6}$, and delete them from the sixty, leaving fifty-six with which I repeat the same process. In this way I obtain, for example, the four symmetric sequences:

5 4 2 1 3 2 1 3 1 2 4 3 1 2 4 3
5 3 4 2 1 3 1 2 3 1 2 4 5 7 8 6

4 5 7 8 6 7 8 6 8 7 5 6 8 7 5 6
5 3 4 2 1 3 1 2 3 1 2 4 5 7 8 6

5 4 2 1 3 2 1 3 1 2 4 3 1 2 4 3
4 6 5 7 8 6 8 7 6 8 7 5 4 2 1 3

4 5 7 8 6 7 8 6 8 7 5 6 8 7 5 6
4 6 5 7 8 6 8 7 6 8 7 5 4 2 1 3.

*   *   *   *   *

(11.) Of the four sequences in *Section 9*, the first may be joined to the fourth, and consequently the second to the third. After this juxtaposition we have two sequences:

5 4 2 1 3 2 1 3 1 2 4 3 1 2 4 3 4 5 7 8 6 7 8 6 8 7 5 6 8 7 5 6
5 3 4 2 1 3 1 2 3 1 2 4 5 7 8 6 4 6 5 7 8 6 8 7 6 8 7 5 4 2 1 3

4 5 7 8 6 7 8 6 8 7 5 6 8 7 5 6 5 4 2 1 3 2 1 3 1 2 4 3 1 2 4 3
5 3 4 2 1 3 1 2 3 1 2 4 5 7 8 6 4 6 5 7 8 6 8 7 6 8 7 5 4 2 1 3.

which are still symmetric, and re-entrant.

In order to join them, it is necessary to destroy the symmetry slightly; but if, for example, we interpolate the second sequence between the terms $\frac{2}{4}$ and $\frac{1}{2}$ of the first, then the whole sequence remains re-entrant, and can therefore start from whichever square one wishes. See figure [2.5], which depicts the *knight's path determined by this sequence.*

*   *   *   *   *

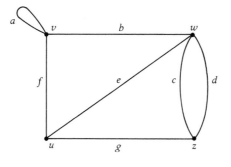

FIG. 2.6.

The knight's-tour problem is a special case of a quite general problem about graphs, namely, the question of when it is possible, in a given graph, to find a circuit which passes through each vertex just once. To be precise, a **circuit** is a path with the additional property that all of its vertices and all of its edges are distinct, with the exception of the first and last vertices, which must coincide. In the graph depicted in Fig. 2.6, the path $v, b, w, c, z, g, u, f, v$ is an example of a circuit. (We remark in passing that this graph exhibits two rather special kinds of circuit: the circuit v, a, v consisting of the edge a joining v to itself, which is a loop; and the circuit w, c, z, d, w containing the two edges c and d which both join w and z—these two edges are said to be **multiple edges**. Many of the graphs we discuss here have neither loops nor multiple edges, and are said to be **simple**).

In the particular case of the knight's-tour problem, the corresponding graph has sixty-four vertices, one for each square of the chessboard, and two vertices are joined by an edge whenever a knight can legally move from one to the other. We are asked to select sixty-four edges which form a circuit passing through each vertex.

The question must be carefully distinguished from the Eulerian problems considered in the previous chapter. In those problems we had to use every edge of the graph in turn, and so we may have to revisit a vertex several times. Here, we use only some of the edges in order to visit each vertex just once. The difference may be compared with the distinction between an explorer and a traveller—the explorer examines all possible routes, whereas the traveller simply wishes to visit each place of interest once. We have already given a simple necessary and sufficient condition for the explorer's problem to have a solution; unfortunately there is no such straightforward test for the traveller's problem.

Kirkman and polyhedra

The first general discussion of the traveller's problem was given by Thomas Penyngton KIRKMAN, who for over fifty years (1840–92) was rector of the small parish of Croft with Southworth in the English county of Lancashire. Despite his isolation and the time devoted to his religious work, he made many contributions to mathematics and was elected a Fellow of the Royal Society in 1857. One of his mathematical interests was the study of polyhedra, that is, solids such as a cube or a pyramid which are bounded by plane faces. The surface of a polyhedron contains the lines of intersection of its faces and the points of intersection of these lines, so that every polyhedron has a corresponding graph whose vertices and edges correspond to these points and lines. A drawing of the graph is often a convenient way of describing the polyhedron; for example, Fig. 2.7 depicts the graphs corresponding to a cube and a pentagonal pyramid. (We shall have more to say about polyhedra and their graphs in Chapter 5.)

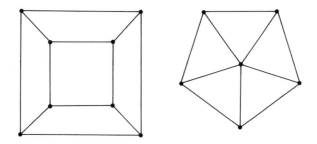

FIG. 2.7.

In a paper [2B] received by the Royal Society on 6 August 1855, Kirkman considered the following question. Given the graph of a polyhedron, can one always find a circuit which passes through each vertex once and only once? He gave a complicated condition (his 'prop.A') which he claimed to be sufficient to ensure that such a circuit can be found, but unfortunately his claim is false. It is possible that he had in mind some additional (even more complicated) restrictions, which he omitted to state explicitly; the addition of such restrictions could easily render the result trivial.

So Kirkman made no serious contribution to that aspect of the problem. His main achievement was to describe a general class of graphs which do not possess a complete circuit of the required kind. This part of his paper is reproduced below; it contains much florid and fanciful terminology which, perhaps fortunately, has not become part of standard mathematical language. To help the reader, we explain that: a 'q-acron' is a polyhedron with q vertices; a 'q-gon' is a circuit with q edges; a 'summit' is a vertex; and any face with a maximum number of edges surrounding it may be designated as the 'base'. Kirkman's example involving 'the cell of the bee' will be described at the end of the extract.

2B

T. P. KIRKMAN
ON THE REPRESENTATION OF POLYEDRA

Philosophical Transactions of the Royal Society, London **146** (1856), 413–418.

*   *   *   *   *

It is easy to prove that there are polyedra on which the closed polygons cannot be drawn.

For suppose the q-gon of prop.A. drawn on a q-acron. In making the circuit of any face G which we enter across an edge FG, which is not an edge of the q-gon, we add to the number of summits counted in F and other faces, all the summits of G, except two, these two having been enumerated in the circuit of F from which we enter G. That is, counting first all the summits of the base, we add to these for every m-gon whose circuit we proceed to make, $m-2$ summits more. The number of faces, connected with each other and with the base by edges, not part of the closed q-gon, whose circuits the closed q-gon makes and includes, will be $\alpha_3 + \alpha_4 + \alpha_5 + \ldots + \alpha_k$; where α_m is the number of m-gons among them and α_k that of the k-gons, the base being one of these. The whole number of summits will therefore be

$$\alpha_3 + 2\alpha_4 + 3\alpha_5 + \ldots + (k-2)\alpha_k + 2 = q;$$

for we have counted *all the summits* of one k-gon, viz. of the base.

In order, then, that such a q-gon should be possible, it is necessary that among the p faces of our p-edral q-acron, there should be α_3 triangles, α_4 quadrilaterals, α_5 pentagons, &c., of which the above equation can be affirmed. Now if q should be odd, and all the p faces even-angled, this equation becomes

$$2\alpha_4 + 4\alpha_6 + 6\alpha_8 + \&c. = 2r+1,$$

which is impossible. Hence it appears, that if the number of summits of a q-acron be odd, while the faces are all even-angled, the closed q-gon cannot be drawn through its summits. I find exceedingly few polyedra on which the closed p-gon and q-gon cannot be drawn. In fact, it is far from being necessary to their existence, that all the conditions of the theorems A and B should be fulfilled.

If we cut in two the cell of the bee by a section of its six parallel edges, we have a 13-acron, whose faces are one hexagon and nine quadrilaterals. The closed 13-gon cannot be drawn. But if a line be drawn from the triedral vertex to the opposite angle of one of the quadrilaterals about that vertex, and this quadrilateral supposed broken into two triangles having that line for their common edge, we shall then have a 13-acron whose faces are one hexagon, eight quadrilaterals, and two triangles; and whose summits are nine *triaces* and [four] *tessaraces*. Of this figure the paradigm can be constructed. Here I would fain beg the reader's permission to call a 5-edral summit a *pentace*, a 6- or 7-edral summit a *hexace* or a *heptace*. The words are at least convenient in speaking of the summits of polyacra.

As authorities and analogy are alike divided about the spelling of the word polyedron, I have pleased myself herein. Why *polyhedron* of necessity, and yet not *perihodic*?

The graph of the polyhedron constructed by Kirkman in his penultimate paragraph is shown in Fig. 2.8; the extra edge he mentioned would join the vertices marked A and B. Kirkman gave a general proof of the fact that, if a polyhedron has an odd number of vertices and each face has an even number of edges, then there is no circuit which passes through all the vertices. The graph in Fig. 2.8 is an example of this.

In fact there is a simpler and more comprehensive way to reach the same conclusion. A graph is said to be **bipartite** if its vertices can be partitioned into two disjoint sets V_1 and V_2 in such a way that each edge joins a vertex in V_1 to a vertex in V_2. Kirkman's graph is bipartite; in Fig. 2.8, we have denoted the vertices in V_1 by circles and those in V_2 by

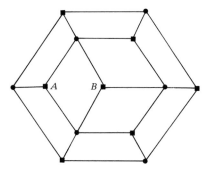

FIG. 2.8.

squares. Since a path in a bipartite graph must visit V_1 and V_2 alternately, it follows that a circuit which begins and ends at the same vertex must contain an even number of vertices. Consequently, if a bipartite graph has an odd number of vertices then it cannot contain a circuit which passes just once through every one of them. This result contains Kirkman's, since the graph of a polyhedron, all of whose faces have an even number of edges, is bipartite.

A special case of the preceding argument was used by Euler [3, p. 332] in a discussion of the knight's tour on a 'non-standard' chessboard—one with n squares in each direction, instead of the usual eight. The chessboard graph, with edges representing knight's moves, is bipartite, since the knight must always move from a white square to a black one, or vice versa. So if $n = 5$, for example, then the total number of squares is odd and the knight cannot make a closed tour of the board.

The Icosian Game

It sometimes happens that the same mathematical idea is discovered by two people independently, at about the same time. It was Kirkman's misfortune to be upstaged by another, more famous, mathematician, William Rowan HAMILTON.

Hamilton was appointed Astronomer Royal of Ireland at the age of twenty-two, was knighted at thirty, and was generally recognized as a leading mathematician of his time. One of his most significant discoveries was the existence of non-commutative algebra, that is, algebra in which multiplication does not necessarily satisfy the rule $xy = yx$. There are many systems of non-commutative algebra, and one of these, which Hamilton discovered and named 'The Icosian Calculus', can be interpreted in terms of paths on the graph of the regular dodecahedron (see below). Hamilton first communicated this discovery in a letter dated 7 October 1856 [4, p. 71], and subsequently developed it in published articles [5], [6]. Furthermore, he used the graphical interpretation as the basis for a puzzle, which he called 'The Icosian Game' (see Frontispiece). He exhibited this game, of which he was inordinately proud, at the meeting of the British Association in Dublin in 1857, and sold the idea for £25 to a wholesale dealer in games and puzzles. It was a bad bargain—for the dealer.

The game was marketed in 1859, accompanied by a printed leaflet [2C] of instructions, written mainly by Hamilton himself. The reader will see that the object of the game was to find paths and circuits on the dodecahedral graph, satisfying certain specified conditions. In particular, the first problem was that of finding a circuit passing just once through each vertex of the graph.

2C

THE ICOSIAN GAME.

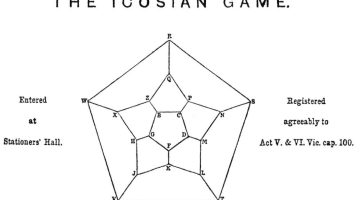

Entered

at

Stationers' Hall.

Registered

agreeably to

Act V. & VI. Vic. cap. 100.

LONDON:

PUBLISHED AND SOLD WHOLESALE BY JOHN JAQUES AND SON, 102 HATTON GARDEN;

AND TO BE HAD AT MOST OF THE LEADING FANCY REPOSITORIES

THROUGHOUT THE KINGDOM,

In this new Game (invented by Sir WILLIAM ROWAN HAMILTON, LL.D., &c., of Dublin, and by him named *Icosian,* from a Greek word signifying 'twenty') a player is to place the whole or part of a set of twenty numbered pieces or men upon the points or in the holes of a board, represented by the diagram above drawn, in such a manner as always to proceed *along the lines* of the figure, and also to fulfil certain *other* conditions, which may in various ways be assigned by another player. Ingenuity and skill may thus be exercised in *proposing* as well as in *resolving* problems of the game. For example, the first of the two players may place the first five pieces in any five consecutive holes, and then require the second player to place the remaining fifteen men consecutively in such a manner that the succession may be *cyclical,* that is, so that No. 20 may be adjacent to No. 1; and it is always possible to answer any question of this kind. Thus, if B C D F G be the five given initial points, it is allowed to complete the succession by following the alphabetical order of the twenty consonants, as suggested by the diagram itself; but after placing the piece No. 6 in the hole H, as before, it is *also* allowed (by the supposed conditions) to put No. 7 in X instead of J, and then to conclude with the succession, W R S T V J K L M N P Q Z. Other Examples of Icosian Problems, with solutions of some of them, will be found in the following page.

EXAMPLES OF ICOSIAN PROBLEMS

FIRST PROBLEM

Five initial points are given; cover the board, and finish *cyclically*. (As hinted in the preceding page, a succession is said to be *cyclical* when the *last* piece is adjacent to the *first*.)

[This problem is always possible in at least two, and sometimes in four different ways. Two examples have been assigned: the following are a few others.]

Example 3. Given B C P N M as initial: two solutions exist; one is the succession, D F K L T S R Q Z X W V J H G; the other is D F G H X W V J K L T S R Q Z.

Example 4. Five initials, L T S R Q. Four solutions.
       "   5. Five initials, J V T S R. Two solutions.

SECOND PROBLEM

Three initial points are given; cover the board *non*-cyclically, and *end* at a given point. (A succession is said to be *non*-cyclical when the last piece is *not* adjacent to the first.)

[This problem is sometimes soluble in only *one* way; sometimes in *two* ways; sometimes in *four* ways; and sometimes it is not soluble at all, as will be seen by the following Examples.]

Example 6. Three initial points, B C D; cover, and end with T. There is in this case only one solution, namely, F G H X Z Q P N M L K J V W R S T.

Example 7. Same initials; cover, and end with W. Two solutions.
       "   8. Same initials; cover, and end with J. Two solutions.

[The same number of solutions exists, if it be required, having the same three initials, to end with K, or L, or N, or V.]

Example 9. Same initials; cover, and end with R. Four solutions.
       "   10. Same initials; cover the board, and end with M. Impossible.

[The same result, if it were required to end with F, or H, or P, or Q, or S, or X.]

THIRD PROBLEM

Certain initial points being given, to *finish* with a given number of additional men. (A succession is said to be *finished*, when no additional piece can, consistently with the rules, be placed upon the board.)

Example 11. Given the four initial points, B C D M; finish with piece No. 10. Solution: L K J H G F.

Example 12. Given the four initials, T V W X; finish with No. 10. One solution.
       "   13. Given the five initials, S N P Q R; finish cyclically with No. 16. One solution.

FOURTH PROBLEM

A given hole being first *stopped* (as by a piece *without* a number), to cover the *rest* of the board, in compliance with proposed conditions.

Example 14. Let B C D be given as the three first points, and let P be stopped: it is required to cover all the other holes, and to end by placing the piece No. 19 at F. Two solutions.

Example 15. Let B C D be still initial, but let L be the stopped hole: to cover the board, and end by placing No. 19 in a hole not adjacent to L. Two solutions.

HINTS ON THE ICOSIAN CALCULUS,

OF WHICH THE ICOSIAN GAME IS DESIGNED TO BE AN ILLUSTRATION.

I.   In a "MEMORANDUM respecting a New System of Roots of Unity," which appeared in the *Philosophical Magazine* for December 1856, Sir W. R. Hamilton expressed himself nearly as follows (a few words only being here omitted):

'I have lately been led to the conception of a new system, or rather *family of systems*, of *non-commutative roots of unity*, which are entirely distinct from the *i j k* of the quaternions, though having some general analogy thereto; and which admit, even more easily than the quaternion symbols do, of geometrical interpretation. In the system which seems at present to be the most interesting one among those included in this new family, I assume three symbols, ι, κ, λ, such that $\iota^2 = 1$, $\kappa^3 = 1$, $\lambda^5 = 1$, $\lambda = \iota\kappa$; where $\iota\kappa$ must be *distinguished* from $\kappa\iota$, since otherwise we should have $\lambda^6 = 1$, $\lambda = 1$. As a very simple *specimen* of the symbolical conclusions deduced from these fundamental assumptions, I may mention that if we make $\mu = \iota\kappa^2 = \lambda\iota\lambda$, we shall have also $\mu^5 = 1$, $\lambda = \mu\iota\mu$; so that μ is a new fifth root of unity, connected with the former fifth root λ by relations of perfect reciprocity. A long train of such symbolical deductions is found to follow; and every one of the results may be *interpreted* as having reference to the passage from *face to face* (or from corner to corner) of the *icosahedron* (or of the dodecahedron): on which account, I am at present disposed to give the name of the 'Icosian Calculus' to this new system of symbols, and of rules for their operations.'

II.   In a LITHOGRAPH, which was distributed in Section A of the British Association, during its Meeting at Dublin in 1857, Sir W. R. H. pointed out a few other symbolical results of the same kind: especially the equations $\lambda\mu^2\lambda = \mu\lambda\mu$, $\mu\lambda^2\mu = \lambda\mu\lambda$, $\lambda\mu^3\lambda = \mu^2$, $\mu\lambda^3\mu = \lambda^2$; and the formula $(\lambda^3\mu^3(\lambda\mu)^2)^2 = 1$, which serves as a *common mathematical type* for the solution of *all cases* of the First Problem of the Game. He also gave at the same time an oral (and hitherto unprinted) account of his rules of *interpretation* of the principal symbols; which rules, with reference to the present Icosian Diagram (or ICOSION), may be briefly stated as follows:

1.   The operation ι *reverses* (or reads backward) a *line* of the figure; changing, for example, BC to CB.

2.   The operation κ causes a line to *turn* in a particular direction round its final point; changing, for instance, BC to DC.

3.   The operation λ changes a line considered as a *side* of a pentagon to the *following side* thereof, proceeding always *right-handedly* for every pentagon except the large or outer one; thus λ changes BC to CD, but SR to RW.

4.   The operation μ is *contrasted* with λ, and changes a line considered as a side of a *different pentagon*, and in the *opposite order* of rotation, to the consecutive side of that *other* pentagon; thus μ changes BC to CP, and SR to RQ; but it changes also RS to ST, whereas λ would change RS to SN.

5.   The only operations employed in the *game* are those marked λ and μ; but another operation, $\omega = \lambda\mu\lambda\mu\lambda = \mu\lambda\mu\lambda\mu$, having the property that $\omega^2 = 1$, was also mentioned in the Lithograph above referred to; and to complete the present statement of *interpretations*, it may be added that the effect of this operation ω is to change an *edge* of a pentagonal *dodecahedron* to the *opposite edge* of that *solid*; for example, in the diagram, BC to TV.

The foregoing Examples, Hints, and Diagram have been supplied to us by the Inventor.

May 1859.   JAQUES AND SON.

There was another version of Hamilton's game, involving a solid dodecahedron (Fig. 2.9), and known as 'The Traveller's Dodecahedron', or 'A Voyage Round the World'. In this game, the vertices represented twenty important places: Brussels, Canton, Delhi, and so on, ending with Zanzibar. Each vertex was marked by a peg, and a thread could be looped around these pegs to indicate a path or a circuit. A complete circuit, passing through each place once only, was called a 'voyage round the world'.

In modern terminology a circuit which passes through each vertex of a graph exactly once is called a **Hamiltonian circuit,** and the graph itself is said to be **Hamiltonian**. It should be apparent that this terminology is not entirely justified. Hamilton, like Vandermonde over eighty years previously, was concerned with one special case, whereas Kirkman discussed a general question; furthermore, although their work was done independently, Kirkman was clearly the first to publish. Nevertheless, the terminology is now too standard to be changed, and we shall use it in the rest of this book.

It is interesting to note that Hamilton visited Kirkman at Croft Rectory during August 1861; from their subsequent letters [4, p. 136] we may judge their mutual esteem at that time. Kirkman wrote that he wished for 'the good fortune to be nearer to such a mathematician as you', and Hamilton replied that 'It would be very difficult for me to express, without having the air of flattering, how much I admire your mathematical genius and discoveries'. However, Hamilton died in 1865, and by 1881 Kirkman [7] had become convinced that he himself had originated

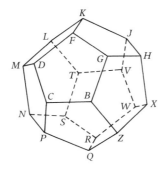

FIG. 2.9.

the discussion of circuits on the dodecahedron, as well as the general treatment of complete circuits. In this delusion he was aided and abetted by Peter Guthrie TAIT [8], a man whose aptitude for causing confusion will be seen again in later chapters of this book. In fact, Kirkman had mentioned the existence of Hamiltonian circuits on the dodecahedron, but not until 1858 [9, p. 160].

Questions of priority are notoriously difficult to assess, since no idea is ever wholly original. In this instance, it is fair to acknowledge that Kirkman inaugurated the general study of complete circuits, while remarking that it was Hamilton's game which generated widespread interest in the problem.

References

1. LEIBNIZ, G. W. *Mathematische Schriften* (1) Vol. 2. Berlin, 1850.
2. GAUSS, C. F. *Werke*, Vol. 5. Göttingen, 1867.
3. EULER, L. Solution d'une question curieuse qui ne paroit soumise à aucune analyse. *Mém. Acad. Sci. Berlin* **15** (1759, published 1766), 310–337 = *Opera Omnia* (1), Vol. 7, 26–56.
4. GRAVES, R. P. *Life of Sir William Rowan Hamilton*, Vol. 3. Dublin University Press, Dublin, 1889.
5. HAMILTON, W. R. Memorandum respecting a new system of roots of unity. *Phil. Mag.* (4) **12** (1856), 446 = *Math. Papers*, Vol. 3, 610.
6. HAMILTON, W. R. [Account of the icosian calculus]. *Proc. Roy. Irish Acad.* **6** (1853–7), 415–416 = *Math. Papers*, Vol. 3, 609.
7. KIRKMAN, T. P. Question 6610, solution by the proposer. *Math. Quest. Solut. Educ. Times.* **35** (1881), 112–116.
8. TAIT, P. G. Listing's *Topologie. Phil. Mag.* (5) **17** (1884), 30–46 = *Sci. Papers.* Vol. 2, 85–98.
9. KIRKMAN, T. P. On the partitions of the *R*-pyramid, being the first class of *R*-gonous *X*-edra. *Phil. Trans. Roy. Soc. London* **148** (1858), 145–161.

3

Trees

A. CAYLEY (1821–95)

IN THE two preceding chapters we have explained how the study of graphs arose from the consideration of various 'recreational' problems, such as the problem of the Königsberg bridges and the knight's tour. The investigation of a special class of graphs, known as 'trees', arose in a rather more mathematical way, from the study of operators in the differential calculus. In this chapter we shall describe some of the early work on the properties of trees, and in the following chapter we shall trace the relationship between this work and some important developments in theoretical chemistry.

Chapters 3 and 4 together survey many of the basic techniques used in the enumeration of graphs with specified properties; consequently, they provide an introduction to the subject known as 'graphical enumeration', an important part of modern graph theory. We shall begin with the work of the nineteenth-century English mathematician Arthur CAYLEY, and end by giving a brief introduction to the fundamental results of George PÓLYA, which were published in the years 1935–1937.

The first studies of trees

A **tree** is defined to be a connected graph which contains no circuits; for example, in Fig. 3.1 we depict all the trees with at most five vertices.

It is easy to deduce some basic properties of trees directly from the definition. In particular, since a tree has no circuits containing only one or two edges, it must be a simple graph according to the definition in

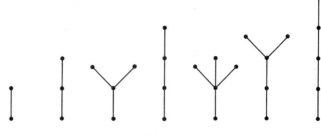

FIG. 3.1.

Chapter 2. Another property, which may be proved by mathematical induction, is that the number of edges is necessarily one less than the number of vertices; conversely, a connected graph with this property must be a tree.

The use of the word 'tree' in this context is presumably derived from the diagrammatic form of these graphs, and is akin to the traditional use of the word in describing genealogical or 'family' trees. The first mathematician to use the term in print was Cayley, in 1857, although the concept had been employed ten years earlier by G. K. C. von Staudt [1], and by the physicist Gustav Robert KIRCHHOFF in his fundamental paper on electrical networks [2]. Cayley pursued the botanical analogy by referring to the vertices of a tree as its 'knots' and to the edges as its 'branches'. Also, in his first investigations he dealt only with trees in which one particular vertex is specially designated as the **root**; this yields what we shall call a **rooted tree** (Cayley later referred to it as a 'root-tree'). For example, from the fourth tree in Fig. 3.1 we may obtain two essentially different rooted trees, depending on whether we choose as the root the central vertex or one of the outer vertices (see Fig. 3.2). A tree with no specified root is sometimes, for emphasis, said to be **unrooted**.

Cayley's first paper on trees [3A] was inspired by some work of his friend James Joseph SYLVESTER on what the latter called 'differential transformation and the reversion of serieses' [3]. Sylvester was well known for his imaginative (and occasionally fantastic) choice of terminology—in fact, as we shall see in the next chapter, he was the first

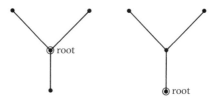

FIG. 3.2.

person to use the word 'graph' in our sense. So it is not impossible that Sylvester was responsible for some of the terminology used in [3A]. The paper began with several paragraphs explaining its relationship with differential operators: these paragraphs are included here mainly for historical reasons, and the reader may be content to skim through them.

Cayley then considered the problem of finding the number A_n of rooted trees with n edges, and he solved this problem by using two elementary techniques and one clever idea. The first technique is that of replacing a sequence of numbers $s_0, s_1, s_2, \ldots$ by a 'generating function'

$$S(x) = s_0 + s_1 x + s_2 x^2 + \ldots,$$

and working with this function, rather than with the sequence. The second technique is based on the observation that if the root, and the edges meeting it, are removed from a rooted tree, then the result is a collection of rooted trees whose roots are the vertices adjacent to the original root. Conversely, any collection of rooted trees may be united to form a single rooted tree, by joining their roots to a new root vertex; for the purpose of this exposition, we shall refer to this process as 'unification'.

Suppose we take one particular rooted tree R with r edges, and list the rooted trees which may be constructed by unification, using copies of R only. First, using no copies of R, we obtain the tree with no edges; then, using one copy of R, we obtain a tree with $r+1$ edges; using two copies of R, we obtain a tree with $2(r+1)$ edges, and so on. So the generating function for the numbers of rooted trees formed in this way is simply

$$1 + x^{r+1} + x^{2(r+1)} + \ldots = (1 - x^{r+1})^{-1}.$$

The result of multiplying together all the generating functions for rooted trees is therefore

$$(1-x)^{-1}(1-x^2)^{-A_1}(1-x^3)^{-A_2}(1-x^4)^{-A_3} \ldots.$$

Cayley's superb insight, allied with his knowledge of the methods used by Euler in the theory of partitions, enabled him to see that the coefficients of the powers of x in this product are formed from the coefficients in the factors in exactly the same way as a rooted tree is constructed by unification. In other words, the above product is equal to

$$1 + A_1 x + A_2 x^2 + A_3 x^3 + \ldots,$$

the generating function for the numbers of rooted trees.

The resulting equation may be used to calculate the numbers A_1, A_2, A_3, ... in turn. Cayley tabulated the first few steps in this calculation, making some errors, some of which he later corrected; we have incorporated these corrections, and also some further ones. The paper ends with a discussion of a related problem involving the enumeration of trees with a given number of 'free branches' (edges which meet vertices of valency one).

A word of warning is necessary before embarking on this paper. Cayley's style was somewhat terse; he tended to be rather free with such phrases as 'it is easy to see that', 'a little consideration will show that', and 'the calculation may be effected very easily'. Furthermore, as the reader will discover, his syntax and punctuation occasionally leave something to be desired.

3A

A. CAYLEY.

ON THE THEORY OF THE ANALYTICAL FORMS CALLED TREES

Philosophical Magazine (4) **13** (1857), 172–176.

A symbol such as $A\partial_x + B\partial_y + \ldots$, where A, B, &c. contain the variables x, y, &c. in respect to which the differentiations are to be performed, partakes of the natures of an operand and operator, and may be therefore called an Operandator. Let P, Q, R, $\ldots$ be any operandators, and let U be a symbol of the same kind, or to fix the ideas, a mere operand; PU denotes the result of the operation P performed on U, and QPU denotes the result of the operation Q performed on PU; and generally in such combinations of symbols, each operation is considered as affecting the operand denoted by means of all the symbols on the right of the operation in question. Now considering the expression QPU, it is easy to see that we may write

$$QPU = (Q \times P)U + (QP)U,$$

where on the right-hand side $(Q \times P)$ and (QP) signify as follows: viz. $Q \times P$ denotes the mere algebraical product of Q and P, while QP (consistently with the general notation as before explained) denotes the result of the operation Q performed upon P as operand; and the two parts $(Q \times P)U$ and $(QP)U$ denote respectively the results of the operations $(Q \times P)$ and (QP) performed each of them upon U as operand. It is proper to remark that $(Q \times P)$ and $(P \times Q)$ have precisely the same meaning, and the symbol may be written in either form indifferently. But without a more convenient notation, it would be difficult to find the corresponding expressions for $RQPU$, &c. This, however, can be at once effected by means of the analytical forms called trees (see figs. [3.3], [3.4], [3.5]), which contain all the trees which can be formed with one branch, two branches, and three branches respectively.

The inspection of these figures will at once show what is meant by the term in question, and by the terms *root, branches* (which may be either main branches, intermediate branches, or free branches), and *knots* (which may be either the root itself, or proper knots, or the extremities of the free branches). To apply this to the question in hand, PU consists of a single term represented by fig. [3.6]; QPU consists, as above, of two terms represented by the two parts of fig. [3.7], viz. the first part represents the term $(Q \times P)U$, and the second part represents the term

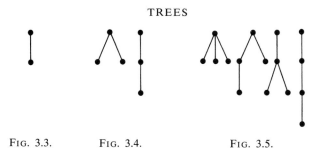

FIG. 3.3.   FIG. 3.4.   FIG. 3.5.

$(QP)U$. And it is obvious that fig. [3.7] is at once formed from the figure [3.6] by adding on a branch terminated by Q at each of the knots of the single part of fig. [3.6]. In like manner $RQPU$ consists of six terms represented by the six parts of fig. [3.8], and this figure is at once formed from fig. [3.7] by adding on a branch terminated by R at each knot of each part of fig. [3.7]. It is hardly necessary to remark that the first part of fig. [3.8] denotes what, in the notation first explained, would be denoted by $(R \times Q \times P)U$, the second term what would in like manner be denoted by $(RQ \times P)U$, and so on, the last part being the term which would be denoted by $((RQ)P)U$; viz. R operates upon Q, giving the operandator RQ, which operates upon P, giving the operandator $(RQ)P$, which finally operates upon U.

The figures [3.6], [3.7], &c. contain the same trees as are contained in the corresponding figures [3.3], [3.4], &c.; only, on account of the different modes of filling up, trees are considered as so many distinct trees in a figure of the second set which are considered as one and the same tree in the corresponding figure of the first set. A difference in the number of trees first occurs in the figures [3.5] and [3.8], the first of which contains only four, while the latter contains six trees, viz. the first tree, the second, third and fourth trees, the fifth tree and the sixth tree of fig. [3.8] correspond respectively to the first tree, the second tree, the third tree, and the fourth tree of fig. [3.5]. To derive fig. [3.8] from fig. [3.5], we must fill up the trees of fig. [3.5] with U at the root and R, Q, P at the other knots in every possible manner, subject only to the restriction, that, reckoning up from the extremity of a free branch to the root, there must not be any transposition in the order of the symbols RQP, and taking care to admit only distinct trees. Thus the first tree of fig. [3.5] might be filled up in six ways; but the trees so obtained are considered as one and the same tree, and we have only the first tree of fig. [3.8]. Again, on account of the restriction, the fourth tree of fig. [3.5] can be filled up in one way only, and we have thus the sixth tree of fig. [3.8]. And thus, in general, each figure of the second set can be formed at once from the corresponding figure of the first set; or when the first set of figures is given, the expression for

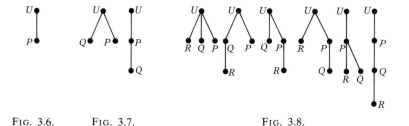

FIG. 3.6.   FIG. 3.7.   FIG. 3.8.

YX...QPU can be formed directly without the assistance of the expression for the preceding symbol *X...QPU*; the number of terms for the *n*th figure of the second set is obviously 1.2.3...*n*, and consequently it is only necessary to count the terms in order to ascertain that no admissible mode of filling up has been omitted.

The number of parts in any one of the figures of the first set is much smaller than the number of parts in the corresponding figure of the second set; and the

$A_1 =$	1 (1)	1	1	1	1	1	1	1	1	1	$(1-x^2)^{-1}$
		1	1	1	1	1	1	1	1	1	
			1	1	1	1	1	1	1	1	
					1	1	1	1	1	1	
							1	1	1	1	
										1	
$A_2 =$	1 1	(2)	2	3	3	4	4	5	5	6	$(1-x^3)^{-2}$
			2	2	4	4	6	6	8	8	
						3	3	6	6	9	
									4	4	
$A_3 =$	1 1	2	(4)	5	7	11	13	17	23	27	$(1-x^4)^{-4}$
				4	4	8	16	20	28	44	
								10	10	20	
$A_4 =$	1 1	2	4	(9)	11	19	29	47	61	91	$(1-x^5)^{-9}$
					9	9	18	36	81	99	
										45	
$A_5 =$	1 1	2	4	9	(20)	28	47	83	142	235	$(1-x^6)^{-20}$
						20	20	40	80	180	
$A_6 =$	1 1	2	4	9	20	(48)	67	123	222	415	$(1-x^7)^{-48}$
							48	48	96	192	
$A_7 =$	1 1	2	4	9	20	48	(115)	171	318	607	$(1-x^8)^{-115}$
								115	115	230	
$A_8 =$	1 1	2	4	9	20	48	115	(286)	433	837	$(1-x^9)^{-286}$
									286	286	
$A_9 =$	1 1	2	4	9	20	48	115	286	(719)	1123	$(1-x^{10})^{-719}$
										719	
$A_r =$	1	2	4	9	20	48	115	286	719	1842	
for $r =$	1	2	3	4	5	6	7	8	9	10	

law for the number of parts, i.e. for the number A_n of the trees with n branches, is a very singular one. To obtain this law, we must consider how the trees with n branches can be formed by means of those of a smaller number of branches. A tree with n branches has either a single main branch, or else two main branches, three main branches, &c. . . . to n main branches. If the tree has one main branch, it can only be formed by adding on to this main branch a tree with $(n-1)$ branches, i.e. A_n contains a part A_{n-1}. If the tree has two main branches, then $p+q$ being a partition of $n-2$, the tree can be formed by adding on to one main branch a tree of p branches, and to the other main branch a tree of q branches; the number of trees so obtained is $A_p A_q$: this, however, assumes that the parts p and q are unequal; if they are equal, it is easy to see that the number of trees is only $\frac{1}{2}A_p(A_p+1)$. Hence $p+q$ being any partition of $n-2$, A_n contains the part $A_p A_q$ if p and q are unequal, and the part $\frac{1}{2}A_p(A_p+1)$ if p and q are equal. In like manner, considering the trees with three main branches, then if $p+q+r$ is any partition of $n-3$, A_n contains the part $A_p A_q A_r$, if p, q, r are unequal; but if two of these numbers, e.g. p and q, are equal, then the part $\frac{1}{2}A_p(A_p+1)A_r$; and if p, q, r are all equal, then the part $\frac{1}{6}A_p(A_p+1)(A_p+2)$; and so on, until lastly we have a single tree with n main branches, or A_n contains the part unity. A little consideration will show that the preceding rule for the formation of the number A_n is completely expressed by the equation

$$(1-x)^{-1}(1-x^2)^{-A_1}(1-x^3)^{-A_2}(1-x^4)^{-A_3}\ldots = 1+A_1 x+A_2 x^2+A_3 x^3+A_4 x^4+\&c.,$$

and consequently that we may, by means of this equation, calculate successively for the different values of n the number A_n of the trees with n branches. The calculation may be effected very easily [as shown in the table opposite.]

I have had occasion, for another purpose, to consider the question of finding the number of trees with a given number of free branches, bifurcations at least. Thus when the number of free branches is three, the trees of the form in question

are those in the annexed figure, and the number is therefore two. It is not difficult to see that we have in this case (B_r being the number of such trees with r free branches),

$$(1-x)^{-1}(1-x^2)^{-B_2}(1-x^3)^{-B_3}(1-x^4)^{-B_4}\ldots = 1+x+2B_2 x^2+2B_3 x^3+2B_4 x^4+\&c.;$$

and a like process of development gives:

$B_r =$	1	2	5	12	33	90
for $r =$	2	3	4	5	6	7

I may mention, in conclusion, that I was led to the consideration of the foregoing theory of trees by Professor Sylvester's researches on the change of the independent variables in the differential calculus.

In 1859 Cayley wrote another paper concerned with the enumeration of a special class of trees [4]. Roughly speaking, he considered rooted trees in which each of the free branches is at the same distance from the root. He showed that if p_n denotes the number of such trees having n free branches, then the 'exponential' generating function P, defined by

$$P(x) = p_1 + p_2 \frac{x}{1!} + p_3 \frac{x^2}{2!} + p_4 \frac{x^3}{3!} + \dots,$$

satisfies the equation

$$e^x P(x) = 2P(x) - 1.$$

This simple equation may be solved for $P(x)$ and the result yields the coefficients p_n in terms of the familiar expansion of e^x.

Despite his expertise in the techniques of enumeration, it was many years before Cayley succeeded in solving the much harder problem of counting unrooted trees—not until 1875, in fact. Before we discuss that work, we shall describe a different approach to the study of trees, due to the French mathematician Camille JORDAN in 1869 [3B].

Jordan considered 'assemblages' of lines intersecting at vertices, and he used the term 'continuous' to signify that such an assemblage was connected. He then defined the 'degree of continuity' of a continuous assemblage to be $n + 1$, where n is the largest number of lines which can be removed without disconnecting the system; in this terminology, a tree is just an assemblage whose degree of continuity is one. Jordan proposed to study connected graphs systematically, by discussing them in order of their degree of continuity, beginning with trees, which have the least continuity.

He began his paper by considering the important question of when two graphs may be regarded as being the same. He observed that apparently different graphs may be *pareil* (translated here as 'similar'); that is, they may be indistinguishable except for the names of their vertices and edges. In modern terminology, two graphs are said to be **isomorphic** when there is a one-to-one correspondence between their vertices, and a one-to-one correspondence between their edges, with the property that corresponding edges join corresponding vertices. An example is shown in Fig. 3.9, where the vertex correspondence $1 \leftrightarrow a$, $2 \leftrightarrow b$, $3 \leftrightarrow c$, $4 \leftrightarrow d$, and the edge correspondence $\alpha \leftrightarrow A$, $\beta \leftrightarrow B$, $\gamma \leftrightarrow C$, $\delta \leftrightarrow D$, $\varepsilon \leftrightarrow E$ establish that the two graphs are isomorphic. The graphs in this example are simple (they have no loops or multiple edges) and in this situation it is clear that

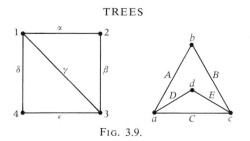

FIG. 3.9.

the edge correspondence is determined by the vertex correspondence, since there is at most one edge joining any pair of vertices. Consequently, in the case of simple graphs, an **isomorphism** may be defined as a vertex correspondence which preserves edges.

It may happen that a graph is isomorphic with itself, in the sense that a relabelling of its vertices gives rise to an isomorphic graph. For example, the relabelling $1 \rightarrow 3$, $2 \rightarrow 4$, $3 \rightarrow 1$, $4 \rightarrow 2$ of the vertices of the first graph in Fig. 3.9 is of this kind. An isomorphism of a graph with itself is usually called an **automorphism** of the graph. (In the case of a graph with loops or multiple edges, the definition is slightly more complicated, since a specific relabelling of the edges must be mentioned.) Of course, every graph has the trivial automorphism which leaves all the labels of the vertices unchanged, but there are many interesting graphs which have non-trivial automorphisms. In the case of the complete graph K_n, every permutation of the vertices is an automorphism, so that there are $n!$ automorphisms altogether. Jordan defined the 'order of symmetry' of a graph to be the number of different automorphisms it has, and he gave the orders of symmetry for various small graphs. He omitted to mention the important fact that the automorphisms of a given graph form a group, although he was one of the first mathematicians to give an account of the theory of groups, and it is possible that the motivation for this paper stemmed from his interest in that subject.

Jordan's main contribution to the theory of trees involved the notion of a 'centre'. He showed that every tree has either one special vertex, or an adjacent pair of special vertices, and that these vertices are, in a specific sense, central. Furthermore, such vertices are useful when studying the automorphisms of a tree, since any automorphism must preserve them.

3B

C. JORDAN
SUR LES ASSEMBLAGES DE LIGNES
[On assemblages of lines]
Journal für die Reine und Angewandte Mathematik **70** (1869), 185–190.

1. Let x, y, z, u, ... be any number of points, and xy, xz, yu, ... straight lines or curves without self-intersections, such that each one joins just two of these points.

We shall call such a system an *assemblage of lines*, whose vertices are x, y, z, u, ... Such an assemblage can be represented symbolically by the quadratic form $xy + xz + yu + \ldots$, whose various terms represent the various lines of the assemblage. Furthermore, if a of these lines have the same ends x, y, then the form representing the assemblage will contain the term xy a times.

Suppose now that xy, xz, ... are those terms of the quadratic form which contains x; yu,... those which have one letter in common with one of the former terms; uv, ... those which have one letter in common with the preceding ones, etc. ... It is clear that the assemblage will be *continuous* if the set of terms xy, xz, ..., yu, ..., uv, ... contains all the letters: if not, not. The assemblage will be said to be $(n+1)$-*fold continuous* if it is possible to remove n of its lines without destroying its continuity.

Two assemblages A and A' will be called *similar* if one can set up a correspondence between their vertices and lines such that to each line xy in one assemblage there corresponds a line $x'y'$ in the other assemblage, where the vertices x' and y' correspond respectively to x and y. In this case, if $axy + bxz + \ldots$ is the form representing A, then $ax'y' + bx'z' + \ldots$ will evidently be the form representing A'.

An assemblage can be similar to itself from several points of view. In this case, A' will be identified with A, and x', y', z', ... will be, in some order, identical with x, y, z, ... Furthermore, the two forms $axy + bxz + \ldots$ and $ax'y' + bx'z' + \ldots$, representing the same network, will necessarily be equal. Thus the substitution which replaces x, y, z, ... by x', y', z', ... does not change the form $axy + bxz + \ldots$. Conversely, let S be a substitution which leaves the form unchanged, and suppose that it replaces x, y, z, ... by x', y', z', ... Let us assign to the vertices x, y, z, ... the vertices x', y', z', ..., and to the a lines joining x to y the corresponding lines joining the corresponding vertices x', y', these lines being taken in any order; and so on ...; then the assemblage will be similar to itself.

The *order of symmetry* of the assemblage, that is the number of ways in which it is similar to itself, is thus equal to $1.2\ldots a.1.2\ldots b\ldots N$, where N is the number of substitutions like S.

2. Problem: *What are the continuous assemblages, not similar to each other, which can be constructed with m lines? What are the degree of continuity and the order of symmetry of each one?*

The following table gives the solution of this problem for $m = 2$, 3, or 4. Each assemblage is described by its representative form. (The corresponding geometrical figure is easily constructed.) We indicate the degree of continuity in Roman numerals, and the order of symmetry in Arabic numerals.

$m = 2$:	$\{\ ab + ac$	I; 2	$2ab$	II; 4
	$\left\{\begin{array}{l} ab + bc + cd \\ ab + bc + bd \\ ab + bc + ca \end{array}\right.$	I; 2	$2ab + bc$	II; 2
$m = 3$:		I; 6	$3ab$	III; 12
		II; 6		
	$\left\{\begin{array}{l} ab + bc + cd + de \\ ab + bc + cd + ce \\ ab + bc + bd + be \\ ab + bc + cd + bd \\ ab + bc + cd + da \\ 2ab + bc + cd \end{array}\right.$	I; 2	$2ab + bc + bd$	II; 4
		I; 2	$2ab + bc + ad$	II; 4
$m = 4$:		I; 24	$2ab + bc + ca$	III; 4
		II; 2	$2ab + 2bc$	III; 8
		II; 8	$3ab + bc$	III; 6
		II; 2	$4ab$	IV; 48

One could extend this table without difficulty to the case $m = 5$, and so on ...,
but the difficulties quickly accumulate. It is easier to get general results, for
arbitrary m, by studying in turn assemblages of continuity one, two, three, ...

3. Assemblages with continuity one. Let x be any vertex of the assemblage. One can
pick out various *limbs* in the assemblage, each one comprising one of the lines xy_1,
xy_2, ... meeting at the vertex x and those lines which can be reached from x by means
of that line. Let p_1, p_2, ... denote the numbers of lines contained in the various
limbs, and let q denote the greatest of these numbers. *The assemblage will
contain a vertex for which $q \leqslant \frac{1}{2}m$ or two vertices joined by a line for both of which
$q = \frac{1}{2}(m+1)$. For all other vertices in the assemblage we shall have $q > \frac{1}{2}(m+1)$.*

For, let us suppose that x is chosen in such a way that q attains its minimum,
and, without loss of generality, take $p_1 = q$. Let xy_1, $x'y_1$, $x''y_1$, ... be the lines
meeting at the vertex y_1, and r, r', r'', ... be the numbers of lines which can be
reached from y_1 by means of these lines: clearly we have

$$r = m - q + 1, \qquad r' + r'' + \ldots = q - 1.$$

Thus $q \leqslant \frac{1}{2}(m+1)$, since otherwise each of r, r', r'', ... would be strictly less than q,
contrary to hypothesis.

Having proved this, suppose first that $q < \frac{1}{2}(m+1)$. Let u be any vertex other
than x; xy_μ, ..., zu the successive lines (k in number) by which u can be reached
from x. The number of lines which are reached from u via the line zu is clearly at
least $k + m - p_\mu$, which is greater than $\frac{1}{2}(m+1)$, since k is at least 1 and p_μ is at
most q.

If $q = \frac{1}{2}(m+1)$ the same proof applies, except that we may have both $k = 1$ and
$p_\mu = q = \frac{1}{2}(m+1)$, whence $\mu = 1$, in which case the vertex u is just y_1.

$$*\qquad*\qquad*\qquad*\qquad*$$

In the case when a tree has a single vertex satisfying Jordan's property
'$q \leqslant \frac{1}{2}m$', we shall call this vertex the **centroid** of the tree; in the case when
there are two vertices satisfying '$q = \frac{1}{2}(m+1)$', we shall refer to them as
the **bicentroid** of the tree. Every tree has either a centroid or a bicentroid,
but not both.

In a subsequent section of his paper Jordan noticed a different procedure which leads to a similar conclusion. For a given tree T, let T' denote
the tree which results when all the free branches are removed from T;
repeating this process yields a sequence of trees T, T', T'', ..., which must
eventually reach either the tree with one vertex or the tree with two
vertices. These vertices are called the **centre**, or **bicentre**, of the original
tree T in the respective cases. To see that this construction is essentially
different from the preceding one, we may consider the tree in Fig. 3.10;
its centroid is the vertex A, whereas its centre is the vertex B.

Counting unrooted trees

Jordan's work was mentioned by Sylvester in an address [5] at the
Royal Institution of Great Britain on 23 January 1874. In addition to
stating that he had independently discovered the existence of the centre

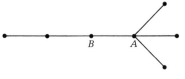

FIG. 3.10.

and bicentre, he drew attention to a link between the study of trees and some contemporary investigations in organic chemistry. Soon afterwards, Cayley published a note [6] on this topic, which was followed by a long paper presented to the meeting of the British Association in 1875 [7]. The latter paper contained a solution of the problem of counting un-rooted trees, as well as some material dealing with applications to chemistry (which we shall discuss in Chapter 4). The key to Cayley's method of enumeration lay in the use of centres and bicentres.

Both Cayley and Sylvester made use of the work of Jordan, as may be verified by reference to the pages of the *Educational Times* for 1877. Since 1862 this journal had published a column of mathematical problems, and among the contributors were many of the leading British mathematicians of the day. Sometimes problems were posed in whimsical terms, and occasionally the whimsy was compounded by being expressed in verse. Some of the more important questions were collected, with additional material, in a half-yearly publication entitled *Mathematical questions, with their solutions, from the "Educational Times"*. This publication is sometimes, misleadingly, referred to as the 'Educational Times Reprints'.

Question 5208 [8] in the mathematical column of the *Educational Times* was set by Sylvester, and it asked for a proof of the existence of Jordan's centroid or bicentroid in a 'ramification'—Sylvester's name for an un-rooted tree. The editors appended some remarks by the author, to the effect that

> ... this remarkable theorem in the pure theory of Colligation, or Cause and Effect, is due to the transcendent genius of M. Camille Jordan; and that it is worthy of notice that an infinite ramification serves to express the possibility of Time, or the natural order of consecution of groups of phenomena, being variable at will, by varying the position of the origin or first cause; and thus greatly extends the common conception of time as a determinate linear order of sequence of such groups.

Although the problem appeared on the first day of April, it is likely that Sylvester's remarks were seriously intended, since he often wrote in a similar vein. There followed a solution by Cayley, using a method rather like that used by Jordan; he also remarked that the centroid and bicentroid are quite distinct from the centre and bicentre.

A few years later, in 1881, Cayley published a much shorter solution of the problem of counting unrooted trees, in which he employed the centroid and bicentroid, rather than the centre and bicentre. This paper is reprinted here as extract [3C]. It will be seen that Cayley distinguished the definitions by referring to them respectively as the 'centre (or bicentre) of length' and the 'centre (or bicentre) of number'. It should also be noted that Cayley here used ϕ_n for what he termed A_{n-1} in [3A].

3C

A. CAYLEY
ON THE ANALYTICAL FORMS CALLED TREES

American Journal of Mathematics **4** (1881), 266–268.

In a tree of N knots, selecting any knot at pleasure as a root, the tree may be regarded as springing from this root, and it is then called a root-tree. The same tree thus presents itself in various forms as a root-tree; and if we consider the different root-trees with N knots, these are not all of them distinct trees. We have thus the two questions, to find the number of root-trees with N knots; and, to find the number of distinct trees with N knots.

I have in my paper "On the Theory of the Analytical Forms called Trees" [3A], given the solution of the first question; viz. if ϕ_N denotes the number of the root-trees with N knots, then the successive numbers ϕ_1, ϕ_2, ϕ_3, etc., are given by the formula

$$\phi_1 + x\phi_2 + x^2\phi_3 + \ldots = (1-x)^{-\phi_1}(1-x^2)^{-\phi_2}(1-x^3)^{-\phi_3}\ldots ,$$

viz. we thus find

suffix of ϕ	1	2	3	4	5	6	7	8	9	10	11	12	13
$\phi =$	1	1	2	4	9	20	48	115	286	719	1842	4766	12486.

And I have in the paper "On the analytical forms called Trees, with application to the theory of chemical combinations" [7], also shown how by the consideration of the centre or bicentre "of length" we can obtain formulae for the number of central and bicentral trees, that is, for the number of distinct trees, with N knots: the numerical result obtained for the total number of distinct trees with N knots is given as follows:

No. of Knots	1	2	3	4	5	6	7	8	9	10	11	12	13
No. of Central Trees	1	0	1	1	2	3	7	12	27	55	127	284	682
'' Bicentral ''	0	1	0	1	1	3	4	11	20	51	108	267	619
Total	1	1	1	2	3	6	11	23	47	106	235	551	1301.

But a more simple solution is obtained by the consideration of the centre or bicentre "of number". A tree of an odd number N of knots has a centre of number, and a tree of an even number N of knots has a centre or else a bicentre of number. To explain this notion (due to M. Camille Jordan) we consider the branches which proceed from any knot, and (excluding always this knot itself) we count the number of the knots upon the several branches; say these numbers are α, β, γ, δ, ε, etc., where of course $\alpha + \beta + \gamma + \delta + \varepsilon + \text{etc.} = N - 1$. If N is even we may have, say $\alpha = \frac{1}{2}N$; and then $\beta + \gamma + \delta + \varepsilon + \text{etc.} = \frac{1}{2}N - 1$, viz. α is larger by unity than the sum of the remaining numbers: the branch with α knots, or the number α, is said to be "merely dominant". If N is odd, we cannot of course have $\alpha = \frac{1}{2}N$, but we may have $\alpha > \frac{1}{2}N$; here α exceeds by 2 at least the sum of the other numbers; and the branch with α knots, or the number α, is said to be "predominant". In every other case, viz. in the case where each number α is less than $\frac{1}{2}N$, (and where consequently the largest number α does not exceed the sum of the remaining numbers), the several branches, or the numbers α, β, γ, etc., are said to be subequal. And we have the theorem. First, when N is odd, there is always one knot (and only one knot) for which the branches are subequal: such knot is called the centre of number. Secondly, when N is even, either there is one knot (and only one knot) for which the branches are subequal: and such knot is then called the centre of number; or else there is no such knot, but there are two adjacent knots (and no other knot) each having a merely-dominant branch: such two knots are called the bicentre of number, and each of them separately is a half-centre.

Considering now the trees with N knots as springing from a centre or a bicentre of number, and writing ψ_N for the whole number of distinct trees with N knots, we readily obtain these in terms of the foregoing numbers ϕ_1, ϕ_2, ϕ_3, etc., viz. we have

$$\psi_1 = 1,$$
$$\psi_2 = \tfrac{1}{2}\phi_1(\phi_1 + 1),$$
$$\psi_3 = \text{coeff. } x^2 \text{ in } (1-x)^{-\phi_1},$$
$$\psi_4 = \tfrac{1}{2}\phi_2(\phi_2 + 1) + \text{coeff. } x^3 \text{ in } (1-x)^{-\phi_1},$$
$$\psi_5 = \text{coeff. } x^4 \text{ in } (1-x)^{-\phi_1}(1-x^2)^{-\phi_2},$$
$$\psi_6 = \tfrac{1}{2}\phi_3(\phi_3 + 1) + \text{coeff. } x^5 \text{ in } (1-x)^{-\phi_1}(1-x^2)^{-\phi_2},$$
$$\psi_7 = \text{coeff. } x^6 \text{ in } (1-x)^{-\phi_1}(1-x^2)^{-\phi_2}(1-x^3)^{-\phi_3},$$

and so on, the law being obvious. And the formulae are at once seen to be true. Thus for $N = 6$, the formula is

$$\psi_6 = \tfrac{1}{2}\phi_3(\phi_3 + 1) + \tfrac{1}{2}\phi_2(\phi_2 + 1).\phi_1 + \phi_2.\tfrac{1}{6}\phi_1(\phi_1 + 1)(\phi_1 + 2) + \tfrac{1}{120}\phi_1(\phi_1 + 1)(\phi_1 + 2)(\phi_1 + 3)(\phi_1 + 4).$$

We have ϕ_3 root-trees with 3 knots, and by simply joining together any two of them, treating the two roots as a bicentre, we have all the bicentral trees with 6 knots: this accounts for the term $\tfrac{1}{2}\phi_3(\phi_3 + 1)$. Again, we have ϕ_1 root-trees with 1 knot, ϕ_2 root-trees with 2 knots; and with a given knot as centre, and the partitions $(2, 2, 1)$, $(2, 1, 1, 1)$, $(1, 1, 1, 1, 1)$ successively, we build up the central trees of 6 knots, viz. (1) we take as branches any two ϕ_2's and any one ϕ_1; (2) any one ϕ_2 and any three ϕ_1's; (3) any five ϕ_1's; the partitions in question being all the partitions of 5 with no part greater than 2, that is, all the partitions with subequal parts. We easily obtain

suffix of ψ	1	2	3	4	5	6	7	8	9	10	11	12	13
$\psi =$	1	1	1	2	3	6	11	23	47	106	235	551	1301

agreeing with the results obtained by the much more complicated formulae of the paper of 1875.

Counting labelled trees

The last extract in this chapter is concerned with another tree-counting problem. In this problem we are given n vertices, and we are asked to find the number t_n of ways of joining these vertices to form a tree; for example, there are $t_4 = 16$ trees on four given vertices (see Fig. 3.11).

As may be inferred from the example, two trees are now to be regarded as distinct when their edges are different, even though they may be isomorphic as abstract graphs. (In fact, there are only two non-isomorphic kinds of tree shown in Fig. 3.11). We emphasize this distinction by referring to the present problem as that of counting **labelled trees**.

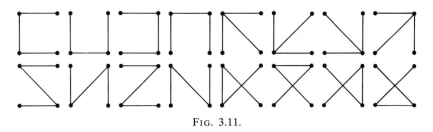

FIG. 3.11.

The solution was first given in 1889, once again by Cayley [9]; his formula for t_n is a very simple one, namely,

$$t_n = n^{n-2}.$$

Calculations equivalent to a proof of this formula had been presented some years earlier by C. W. Borchardt [10, p. 121] and Sylvester [11, p. 55], but neither of these authors was in any way concerned with the problem in the particular guise considered here.

Cayley's proof of the formula for t_n was somewhat less than complete. His paper began as follows:

> The number of trees which can be formed with $n+1$ given knots α, β, γ, ... is $= (n+1)^{n-1}$; for instance $n = 3$, the number of trees with the 4 given knots α, β, γ, δ is $4^2 = 16$, for in the first form shown in the figure the α, β, γ, δ may be arranged in 12 different orders ($\alpha\beta\gamma\delta$ being regarded as equivalent to $\delta\gamma\beta\alpha$), and in the second form any one of the 4 knots α, β, γ, δ may be in the place occupied by the α: the whole number is thus $12+4, = 16$.
>
> Considering for greater clearness a larger value of n, say $n = 5$, I state the particular case of the theorem as follows:
>
> No. of trees $(\alpha, \beta, \gamma, \delta, \varepsilon, \zeta) =$ No. of terms of $(\alpha + \beta + \gamma + \delta + \varepsilon + \zeta)^4 \alpha\beta\gamma\delta\varepsilon\zeta$,
>
> $$= 6^4, = 1296,$$

$$12+4=16$$

and it will be at once seen that the proof given for this particular case is applicable for any value whatever of n.

Nevertheless, it is not at all obvious how the proof which he then proceeded to give for the case $n = 5$ can be extended to larger values of n, and so we shall not reproduce it here. Instead, we shall present an alternative proof given some years later by the German mathematician Heinz PRÜFER [3D]. Prüfer's interest in the problem arose from a problem on permutations; the details of this need not concern us, and the paper can be understood without knowing them. It should be noted that Prüfer was apparently quite unaware of the work of Cayley, and also of the terminology of graph theory, which was by that time (1918) fairly extensive.

The method described by Prüfer depends on the construction of an explicit one-to-one correspondence between the set of sequences $\{a_1, a_2, \ldots, a_{n-2}\}$, in which each a_i is one of the n given vertices, and the set of trees which can be constructed on these vertices. Since there are exactly n^{n-2} sequences of the type just described, the required result follows immediately.

3D

H. PRÜFER

NEUER BEWEIS EINES SATZES ÜBER PERMUTATIONEN

[A new proof of a theorem on permutations]

Archiv der Mathematik und Physik (3) **27** (1918), 142–144.

Herr Dziobek [12] has announced a theorem which can be formulated as follows:

The number of ways of obtaining all possible permutations on n symbols using n − 1 transpositions is exactly equal to n^{n-2}.

Herr Dziobek proved his theorem by setting up a recursion formula for calculating the required number. His proof that the number n^{n-2} satisfies this recursion formula is not particularly simple, and it is perhaps of interest to look at another proof which depends entirely on combinatorial considerations. I shall express it in an intuitive geometrical form, as suggested by Professor Schur in a problem for the Mathematics Seminar at the University of Berlin.

*Consider a country with n towns. These towns must be connected by a railway
network of n − 1 lines (the smallest possible number) in such a way that one can get
from each town to every other town. There are n^{n-2} different railway networks of this
kind.*

By a line, we mean a stretch of railway which connects two towns.

The theorem can be proved by assigning to each railway network, in a unique
way, a symbol $\{a_1, a_2, \ldots, a_{n-2}\}$, whose $n-2$ elements can be selected independently from any of the numbers 1, 2, ..., n. There are n^{n-2} such symbols, and
this fact, together with the one-to-one correspondence between networks and
symbols, will complete the proof.

In the case $n = 2$ we assign the empty symbol to the only possible network
connecting the two towns. If $n > 2$, we denote the towns by the numbers 1, 2, ...,
n and specify them in a fixed sequence. The towns at which only one line
terminates we call the endpoints of the network. In every net there are endpoints.
For otherwise there would be at least two lines terminating at each town; and
there would be at least $\dfrac{2n}{2} = n$ lines.

In order to define, for $n > 2$, the symbol belonging to a given net, one proceeds
as follows:

Let b_1 be the first town which is an endpoint of the net, and a_1 the town which
is directly joined to b_1. Then a_1 is the first element of the symbol. We now strike
out the town b_1 and the line $a_1 b_1$. There remains a net containing $n-2$ lines
which connect $n-1$ towns in such a way that one can get from each town to any
other.

If also $n-1 > 2$, then one determines the town a_2 with which the first endpoint
b_2 of the new net is directly connected. We take a_2 as the next element of the
symbol. Then we strike out the town b_2 and the line $a_2 b_2$. We obtain a net with
$n-3$ lines and the same properties.

We continue this procedure until we finally obtain a net with only one line
joining 2 towns. The symbol is then complete.

For example:

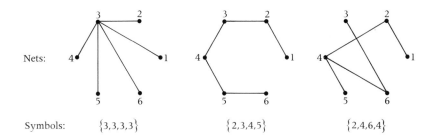

Nets:

Symbols: $\{3,3,3,3\}$ $\{2,3,4,5\}$ $\{2,4,6,4\}$

Each town at which m lines terminate occurs exactly $m-1$ times in the symbol.
For, in the formation of the symbol by successively removing lines, a town
appears in the symbol only when one of its incident edges is removed, except in
the case that this edge is the last one having that town as endpoint.

Conversely, if we are given a particular symbol $\{a_1, a_2, \ldots, a_{n-2}\}$ other than the
empty symbol, then we write down the numbers 1, 2, ..., n and find the first
number which does not appear in the symbol. Let this be b_1. Then we connect the

towns b_1 and a_1 by a line. We now strike out the first element of the symbol and the number b_1.

If $\{a_2, a_3, \ldots, a_{n-2}\}$ is also not the empty symbol, then we find b_2, the first of the $n-1$ remaining numbers which does not appear in the symbol, and connect the towns b_2 and a_2. Then we strike out the number b_2 and the element a_2 in the symbol.

In this way we eventually obtain the empty symbol. When that happens, we join the last two towns not yet crossed out.

That the system of lines obtained by this construction actually is a net, and that this net and no other actually gives rise to the given symbol, follows by an induction argument. For, if a net is represented by a symbol, then the towns which do not appear in the symbol are just the endpoints of the net. As the line $b_1 a_1$ is the only line ending at b_1, it must appear in the net. But we may assume that we have proved that the symbol $\{a_2, a_3, \ldots, a_{n-2}\}$ corresponds to just one net connecting all the towns except b_1, and that this net was obtained by the construction, so that we also get the truth of the proposition for the symbol $\{a_1, a_2, \ldots, a_{n-2}\}$.

References

1. VON STAUDT, G. K. C. *Geometrie der Lage.* Nürnberg, 1847.
2. KIRCHHOFF, G. R. Über die Auflösung der Gleichungen, auf welche man bei der Untersuchung der linearen Vertheilung galvanischer Ströme geführt wird. *Ann. Phys. Chem.* **72** (1847), 497–508 = *Ges. Abh.*, 22–33. [8A]
3. SYLVESTER, J. J. On differential transformation and the reversion of serieses. *Proc. Roy. Soc. London* **7** (1854–5), 219–223 = *Math. Papers*, Vol. 2, 50–54.
4. CAYLEY, A. On the theory of the analytical forms called trees—part II. *Phil. Mag.* (4) **18** (1859), 374–378 = *Math. Papers*, Vol. 4, 112–115.
5. SYLVESTER, J. J. On recent discoveries in mechanical conversion of motion. *Proc. Roy. Inst. Gt. Brit.* **7** (1873–5), 179–198 = *Math. Papers*, Vol. 3, 7–25.
6. CAYLEY, A. On the mathematical theory of isomers. *Phil. Mag.* (4) **47** (1874), 444–446 = *Math. Papers*, Vol. 9, 202–204. [4B]
7. CAYLEY, A. On the analytical forms called trees, with application to the theory of chemical combinations. *Rep. Brit. Assoc. Advance. Sci.* **45** (1875), 257–305 = *Math. Papers*, Vol. 9, 427–460.
8. SYLVESTER, J. J. Question 5208. *Math. Quest. Solut. Educ. Times* **27** (1877), 81.
9. CAYLEY, A. A theorem on trees. *Quart. J. Pure Appl. Math.* **23** (1889), 376–378 = *Math. Papers*, Vol. 13, 26–28.
10. BORCHARDT, C. W. Über eine der Interpolation entsprechende Darstellung der Eliminations-Resultante. *J. Reine Angew. Math.* **57** (1860), 111–121.
11. SYLVESTER, J. J. On the change of systems of independent variables. *Quart. J. Pure Appl. Math.* **1** (1857), 42–56, 126–134 = *Math. Papers*, Vol. 2, 65–85.
12. DZIOBEK, O. Eine Formel der Substitutionstheorie. *Sitzungsber. Berliner Math. Ges.* **17** (1917), 64–67.

4

Chemical Graphs

J. J. SYLVESTER (1814–97)

THE interchange of ideas between two different branches of science is often highly beneficial to both of them. The aim of this chapter is to describe how Cayley related his work on trees to the studies of chemical composition which were being carried on at the same time—in particular, to the investigation of pairs of compounds having the same chemical composition, but different properties. This fusion of mathematical and chemical ideas inspired some of the terminology which is now standard in graph theory, including the word 'graph' itself.

Graphic formulae in chemistry

We shall begin by describing briefly some of the fundamental advances which were made in theoretical chemistry during the nineteenth century. By 1850 it had become clear that chemical compounds are formed from constituent elements, and that these elements combine together in certain fixed proportions. Consequently, chemical formulae, such as H_2O for water and C_2H_5OH for ethyl alcohol, were well established. It was thus natural that chemists should speculate on the mechanisms underlying the laws of chemical combination. There is a well-known account of how inspiration came to the German chemist F.A. Kekulé in 1854, on the top of a London omnibus [1]:

> During my stay in London I lived for a long time in Clapham Road, near to the common. I often spent the evenings with my friend Hugo Müller in Islington, at the opposite end of that vast town. There we talked about

many things but most often of our beloved chemistry. One fine summer day I was riding once again on the last omnibus through the deserted streets of the metropolis, at other times so crowded—'outside', on the roof of the omnibus, as usual. I sank into a day-dream. There before my eyes gambol-led the atoms. I had always seen them in motion, those tiny beings, but I had never succeeded in discerning the nature of their motion. This day I saw how frequently two smaller atoms joined themselves together like a loving couple; how larger ones embraced two smaller ones, how still larger ones seized three and even four of the smaller ones, and how the whole twisted in a whirling dance. I saw how the larger ones formed a line and dragged along still smaller ones at the ends of the chain. I saw what our past master Kopp, my highly respected teacher and friend, described to us in such a delightful manner in his 'Molecularwelt', but I saw it long before him. The cry of the conductor 'Clapham Road' roused me from my musings, but I stayed up part of the night in order to commit to paper at least the outlines of that vision. In this way began the structure theory.

Kekulé was not the only person to think in these terms. Other claims for priority have been made by (or for) several chemists: A. M. But-lerov, a Russian; A. S. Couper, a Scot; and Edward FRANKLAND, who wrote a paper on 'the doctrine of atomicity' in 1852. It is not for us to discuss these rival claims in detail here, but the reader interested in knowing more about them is recommended to refer to the book by C. A. Russell [2].

From the confused dreams and imaginings of Kekulé and his contem-poraries sprang the need to represent chemical compounds by diagrams. In fact, diagrams had been used as early as 1789 by William Higgins [3], who employed figures like $\begin{smallmatrix} I & \!\!\!—\!\!\! & D \\ & \diagdown\diagup & | \\ d & \!\!\!—\!\!\! & S \end{smallmatrix}$, representing ferrous sulphate ($FeSO_4$), where I stands for iron, S stands for sulphur, and D and d stand for dephlogisticated air (oxygen). But Higgins was ahead of his time, and his work had little influence on the development of chemical theory. However, by the middle of the nineteenth century, ideas about atoms and the way they combine had been clarified to the point where meaningful diagrams of chemical molecules could be drawn. The introduction of such diagrams was closely related to the evolution of one of the most important of these new ideas—the concept of valency, which arose from the known facts about chemical equivalents of elements. For example, it was known that in marsh-gas, CH_4, four equivalents of hydrogen com-bined with one equivalent of carbon; this could be explained by supposing that each carbon atom has four 'bonds' by which it may be linked to other atoms, whereas each hydrogen atom has just one bond. Nowadays we express this fact by saying that carbon has valency four and hydrogen has valency one. The use of the same terminology in graph theory is a result of the link between chemistry and graph theory, which will become clear in the course of this chapter.

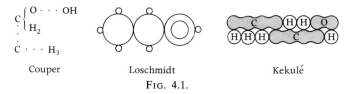

Couper   Loschmidt   Kekulé
FIG. 4.1.

In Fig. 4.1 we illustrate three early attempts to represent molecules diagrammatically; the substance represented is ethyl alcohol, C_2H_5OH. The first diagram is due to Couper, and is typical of the notation he introduced in 1858 [4]. (The presence of two oxygen atoms is due to his mistaken belief that the atomic weight of oxygen was 8, rather than the correct value, 16.) His system had several disadvantages, the most serious of which was the lack of provision for the use of multiple bonds. Nevertheless, his use of dotted lines to represent bonds was a helpful technique, and the resulting system was undoubtedly superior to the other two, due to the Austrian chemist J. Loschmidt and to Kekulé. Loschmidt's system dates from 1861 [5]. He used circles of different sizes to represent different atoms, and indicated the presence of a bond by making the circles touch; multiple bonds were represented by means of overlapping circles. Kekulé's system was introduced in a footnote to the first volume of his *Lehrbuch der organischen Chemie* [6], which also appeared in 1861. Univalent atoms were represented by circles, while atoms of higher valency were shown as 'sausages', with a number of bulges corresponding to the valency.

An important advance occurred in 1864, when the Edinburgh chemist Alexander CRUM BROWN introduced his version of the 'graphic notation' [7]. Each atom was shown separately, represented by a letter enclosed in a circle, and all single and multiple bonds were marked by lines joining the circles (Fig. 4.2). Crum Brown's system is essentially the one in use today, except that the circles are now usually omitted.

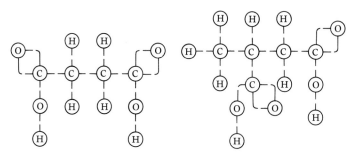

Succinic acid   Pyrotartaric acid

FIG. 4.2.

At first, Crum Brown's work was available only in an Edinburgh journal, but in 1866 his ideas became accessible to a wider readership. In that year Frankland published an introductory text called *Lecture notes for chemical students,* which did much to propagate Crum Brown's system. The first extract [4A] in this chapter is a short section from Frankland's book. After explaining the representation of atoms and bonds by means of circles with lines emanating from them, he went on to describe the graphic notation.

4A

E. FRANKLAND
GRAPHIC NOTATION

Lecture notes for chemical students. London, 1866.

This mode of notation, although far too cumbrous for general use, is invaluable for clearly showing the arrangement of the individual atoms of a chemical compound. It is true that it expresses nothing more than the symbolic notation of the same compound, if the latter be written and understood as above described; nevertheless the graphic form affords most important assistance, both in fixing upon the mind the true meaning of symbolic formulae, and also in making comparatively easy of comprehension the internal arrangement of the very complex molecules frequently met with both in mineral and organic compounds. It is also of especial value in rendering evident the causes of isomerism in organic bodies.

Graphic notation, like the above method of symbolic notation, is founded almost entirely upon the doctrine of atomicity, and consists in representing, graphically, the mode in which every bond in a chemical compound is disposed of. Inasmuch, however, as the principles involved are precisely the same as those already described under the heads of SYMBOLIC NOTATION and ATOMIC-ITY OF ELEMENTS, it is unnecessary here to do more than give the following comparative examples of symbolic and graphic formulae:-

	Symbolic.	Graphic.
Water	OH_2.	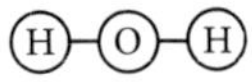 (H)—(O)—(H)
Nitric acid	NO_2Ho.	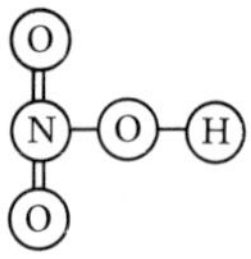 (O)‖(N)—(O)—(H), (O)
Ammonic chloride	NH_4Cl.	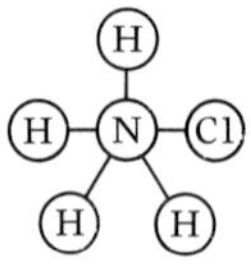 (H), (H)—(N)—(Cl), (H) (H)

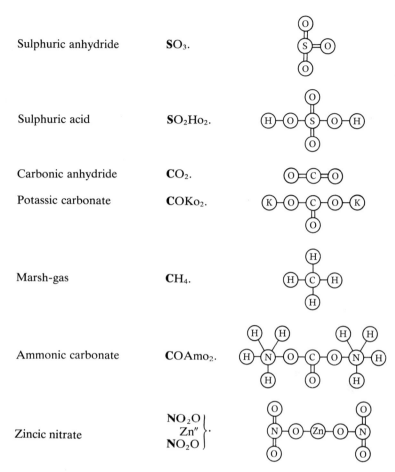

Sulphuric anhydride	SO_3.
Sulphuric acid	SO_2Ho_2.
Carbonic anhydride	CO_2.
Potassic carbonate	$COKo_2$.
Marsh-gas	CH_4.
Ammonic carbonate	$COAmo_2$.
Zincic nitrate	$\left.\begin{array}{c} NO_2O \\ Zn'' \\ NO_2O \end{array}\right\}.$

It must be carefully borne in mind that these graphic formulae are intended to represent neither the shape of the molecules, nor the relative position of the constituent atoms. The lines connecting the different atoms of a compound, and which might with equal propriety be drawn in any other direction, provided they connected together the same elements, serve only to show the definite disposal of the bonds: thus the formula for nitric acid indicates that two of the three constituent atoms of oxygen are combined with nitrogen alone, and are consequently united to that element by both their bonds, whilst the third oxygen atom is combined both with nitrogen and hydrogen.

The lines connecting the different atoms of a compound are but crude symbols of the bonds of union between them; and it is scarcely necessary to remark that no such material connexions exist, the bonds which actually hold together the atoms of a compound being in all probability, as regards their nature, much more like those which connect the members of our solar system.

It may also be here mentioned that graphic, like symbolic formulae, are purely statical representations of chemical compounds, they take no cognizance of the amount of potential energy associated with the different elements. Thus in the formulae for marsh-gas and carbonic anhydride, there is no indication that the molecule of the first compound contains a vast store of force, whilst the last is comparatively a powerless molecule.

Isomerism

Crum Brown's notation was soon accepted everywhere, after some resistance from Kekulé and others. Its acceptance was partly due to its success in explaining one strange experimental fact—that there are pairs of substances which have the same chemical composition, although their physical properties are different. The graphic notation made it plain that this is because the atoms are arranged in different ways in the different

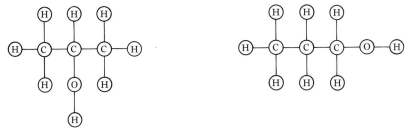

FIG. 4.3.

substances. For example, the original paper of 1864 [7] gave the graphic formulae (Fig. 4.3) for two substances with the composition C_3H_7OH, known to Crum Brown as propylic alcohol and Friedel's alcohol.

This is an instance of what is now called *isomerism*, and the two compounds are known as *isomers;* in many cases there are more than two isomers with the same constitutional formula. By 1874, the phenomenon and its explanation in terms of the graphic notation were widely known, and Cayley wrote a paper [4B] explaining how his earlier work on trees could usefully be applied to the subject.

4B

A. CAYLEY
ON THE MATHEMATICAL THEORY OF ISOMERS

Philosophical Magazine (4) **47** (1874), 444–446.

I consider a "diagram," viz. a set of points *H, O, N, C,* &c. (any number of each), connected by links into a single assemblage under the condition that through each H there passes not more than one link, through each O not more than two links, through each N not more than three links, through each C not more than four links. Of course through every point there passes at least one link, or the points would not be connected into a single assemblage.

In such a diagram each point having its full number of links is saturate, or nilvalent: in particular, each point H is saturate. A point not having its full number of links is univalent, bivalent, or trivalent, according as it wants one, two, or three of its full number of links. If every point is saturate the diagram is saturate, or nilvalent; or, say, it is a "plerogram"; but if the diagram is susceptible of n more links, then it is n-valent; viz. the valency of the diagram is the sum of the valencies of the component points.

Since each H is connected by a single link (and therefore to a point O, C, &c. as the case may be, but not to another point H), we may without breaking up the diagram remove all the points H with the links belonging to them, and thus obtain a diagram without any points H: such a diagram may be termed a "kenogram": the valency is obviously that of the original diagram *plus* the number of removed H's.

If from a kenogram, we remove every point O, C, &c. connected with the rest of the diagram by a single link only (each with the link belonging to it), and so on indefinitely as long as the process is practicable, we arrive at last at a diagram in which every point O, C, &c. is connected with the rest of the diagram by two links at least: this may be called a "mere kenogram."

Each or any point of a mere kenogram may be made the origin of a "ramification"; viz. we have here links branching out from the original point, and then again from the derived points, and so on any number of times, and never again uniting. We can thus from the mere kenogram obtain (in an infinite variety of ways) a diagram. The diagram completely determines the mere kenogram; and consequently two diagrams cannot be identical unless they have the same mere kenogram. Observe that the mere kenogram may evanesce altogether; viz. this will be the case if the diagram or kenogram is a simple ramification.

A ramification of n points C is $(2n+2)$-valent: in fact, this is so in the most simple case $n=1$; and admitting it to be true for any value of n, it is at once seen to be true for the next succeeding value. But no kenogram of points C is so much as $(2n+2)$-valent; for instance, 3 points C linked into a triangle, instead of being 8-valent are only 6-valent. We have therefore plerograms of n points C and $2n+2$ points H, say plerograms C^nH^{2n+2}; and in any such plerogram the kenogram is of necessity a ramification of n points C; viz. the different cases of such ramifications are

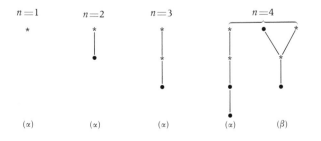

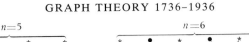

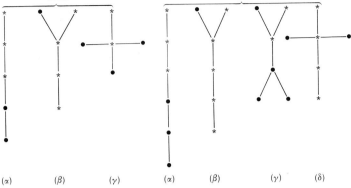

where the mathematical question of the determination of such forms belongs to the class of questions considered in my paper "On the Theory of the Analytical Forms called Trees" [3A], [8], and in some papers on Partitions in the same Journal. (The distinction in the diagrams of asterisks and dots is to be in the first instance disregarded; it is made in reference to what follows, the explanation as to the allotrious points.)

The different forms of univalent diagrams $C^n H^{2n+1}$ are obtained from the same ramifications by adding to each of them all but one of the $2n+2$ points H; that is, by adding to each point C except one its full number of points H, and to the excepted point one less than the full number of points H. The excepted point C must therefore be univalent at least; viz. it cannot be a saturate point, which presents itself for example in the diagrams $n=5(\gamma)$ and $n=6(\delta)$. And in order to count the number of distinct forms (for the diagrams $C^n H^{2n+1}$), we must in each of the above ramifications consider what is the number of distinct classes into which the points group themselves, or, say, the number of "allotrious" points. For instance, in the ramification $n=3$ there are two classes only; viz. a point is either terminal or medial; or, say, the number of allotrious points is $=2$: this is shown in the diagrams by means of the asterisks; so that in each case the points which may be considered allotrious are represented by asterisks, and the number of asterisks is equal to the number of allotrious points.

Thus, number of univalent diagrams $C^n H^{2n+1}$:

$n=1$,		1
$n=2$,		1
$n=3$,		2
$n=4$, $(\alpha)2$; $(\beta)2$;	together	4
$n=5$; $(\alpha)3$; $(\beta)4$; $(\gamma)1$;	together	8
$n=6$, $(\alpha)3$; $(\beta)5$; $(\gamma)2$; $(\delta)3$;	together	13

where it will be observed that, $n=5(\gamma)$, and $n=6(\delta)$, the numbers of allotrious points are 2 and 4 respectively; but since in each of these cases one point is saturate, they give only the numbers 1 and 3 respectively. It might be mathematically possible to obtain a general solution; but there would be little use in this; and for even the next succeeding case, No. of bivalent diagrams $C^n H^{2n}$; the extreme complexity of the question would, it is probable, prevent the attainment of a general solution.

Passing to the chemical signification of the formulae, and instead of the radicals C^nH^{2n+1} considering the corresponding alcohols $C^nH^{2n+1}.OH$, then, $n = 1, 2, 3,$ 4, the numbers of known alcohols are 1, 1, 2, 4, agreeing with the foregoing theoretic number (see [9]); but $n = 4$, the number of known alcohols is $= 2$, instead of the foregoing theoretic number 8. It is, of course, no objection to the theory that the number of theoretic forms should exceed the number of known compounds; the missing ones may be simply unknown; or they may be only capable of existing under conceivable, but unattained, physical conditions (for instance, of temperature); and if defect from the theoretic number of compounds can be thus accounted for, the theory holds good without modification. But it is also possible that the diagrams, in order that they may represent chemical compounds, may be subject to some as yet undetermined conditions; viz. in this case the theory would stand good as far as it goes, but would require modification.

Despite his original opinion that 'there would be little use' in obtaining an expression for the number of diagrams with a given carbon content, Cayley later wrote further papers on this problem. We shall illustrate his work by reference to the series of compounds with constitutional formula C_nH_{2n+2}, for $n = 1, 2, \ldots$; these compounds are known as *paraffins* or *alkanes*.

Cayley's interest was aroused by the fact that the graphic formulae for paraffins (and some other series of compounds) are representations of trees. This is not the case for all molecules, a famous example being the hexagonal formula for benzene discovered by Kekulé. To check the result for the paraffins, we note that each carbon atom has four bonds and each hydrogen one, and every bond has two ends, so that the total number of bonds is

$$\tfrac{1}{2}\{4.n + 1.(2n+2)\} = 3n + 1.$$

Since there are $3n + 2$ atoms in all, and since this is one more than the number of bonds, the graphic formula must be a tree. For a given value of n there may be several trees corresponding to C_nH_{2n+2}; for example, Fig. 4.4 depicts the molecules of butane and isobutane, both having the

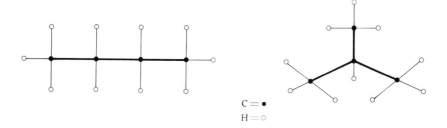

C $= \bullet$
H $= \circ$

FIG. 4.4.

formula C_4H_{10}. By ignoring the hydrogen atoms, Cayley obtained a tree made up entirely of carbon atoms, in which the valency of each atom is at most four. (These 'carbon-trees' are drawn with heavier lines in Fig. 4.4.) This procedure is reversible, since from any carbon-tree we may obtain a paraffin by joining to each carbon atom enough hydrogen atoms to ensure that its valency is four. It follows that the problem of finding the number of different paraffins with the formula C_nH_{2n+2} is essentially the same as the problem of counting unrooted trees with n vertices, where no vertex has valency exceeding four.

In 1875, Cayley discovered a systematic method of counting these carbon-trees. He presented his method in a lengthy paper [10] read to the British Association, and at the same time he outlined a solution of the general problem of counting unrooted trees. The computations involved were laborious, and his paper contained many pages of tabulated calculations. Nevertheless, he did succeed in finding the numbers of carbon-trees with n vertices for values up to $n = 11$. Subsequently he published more papers on this problem [11], [12], but he was unable to obtain a neat or compact solution. Further progress required more sophisticated techniques, which we shall outline in the final section of this chapter. But first, we shall describe how the term 'graph' came into existence.

Clifford, Sylvester, and the term 'graph'

Cayley's long paper on trees and isomerism was not the only paper of graphical significance presented to the British Association at their meeting in Bristol in 1875. In a report [13] of the transactions of the mathematical section we read that:

> Professor Clifford astonished the Section by some remarkable applications of Grassman's "polar multiplication" defined by the law that ba is *minus ab*. He applied it to the graphical representation of invariants, to the expansion of unsymmetrical functions, and to the notation of matrices, illustrating his remarks by drawings representing atoms hung together in various ways.

The astonishing gentleman was William Kingdon CLIFFORD, already (at the age of thirty) a leading figure on the British mathematical scene. The main subject of his talk was the study of the invariants of binary quantics. We do not need to go into the technicalities of this subject here; it is sufficient to say that a binary quantic is an homogeneous expression in two variables, such as $ax^3 + 3bx^2y + 3cxy^2 + dy^3$, and that an invariant is a function of the coefficients a, b, c, d which is unaltered by certain transformations of the variables. Both Cayley and Sylvester had made important contributions to the study of invariants, and Clifford introduced the fashionable graphic notation into the subject.

In 1876 Sylvester crossed the Atlantic to become the first Professor of

Mathematics in the newly-founded Johns Hopkins University in Baltimore. The years he spent in Baltimore were happy ones, and his fertile imagination brought forth many new ideas, some good, and some not so good. In particular, he conceived the notion that there is a relationship between chemistry and algebra, exemplified by the use of graphic notation. He published a note on his ideas in February 1878, in the scientific journal *Nature* [4C]. In this note the word 'graph' is used in our sense for the first time; it is clearly derived from the chemical 'graphic notation'. So the credit (or blame) for the use of this term must be ascribed to Sylvester.

4C

J. J. SYLVESTER
CHEMISTRY AND ALGEBRA

Nature **17** (1877–8), 284.

It may not be wholly without interest to some of the readers of *Nature* to be made acquainted with an analogy that has recently forcibly impressed me between branches of human knowledge apparently so dissimilar as modern chemistry and modern algebra. I have found it of great utility in explaining to non-mathematicians the nature of the investigations which algebraists are at present busily at work upon to make out the so-called *Grundformen* or irreducible forms appurtenant to binary quantics taken singly or in systems, and I have also found that it may be used as an instrument of investigation in purely algebraical inquiries. So much is this the case that I hardly ever take up Dr. Frankland's exceedingly valuable *Notes for Chemical Students*, which are drawn up exclusively on the basis of Kekulé's exquisite conception of *valence*, without deriving suggestions for new researches in the theory of algebraical forms. I will confine myself to a statement of the grounds of the analogy, referring those who may feel an interest in the subject and are desirous for further information about it to a memoir which I have written upon it for the new *American Journal of Pure and Applied Mathematics*, the first number of which will appear early in February.

The analogy is between atoms and *binary* quantics exclusively.

I compare every binary quantic with a chemical atom. The number of factors (or rays, as they may be regarded by an obvious geometrical interpretation) in a binary quantic is the analogue of the number of *bonds*, or the *valence*, as it is termed, of a chemical atom.

Thus a linear form may be regarded as a monad atom, a quadratic form as a duad, a cubic form as a triad, and so on.

An invariant of a system of binary quantics of various degrees is the analogue of a chemical substance composed of atoms of corresponding *valences*.

The order of such invariant in each set of coefficients is the same as the number of atoms of the corresponding *valence* in the chemical compound.

A co-variant is the analogue of an (organic or inorganic) compound radical. The orders in the several sets of coefficients corresponding, as for invariants, to the respective valences of the atoms, the free valence of the compound radical then becomes identical with the degree of the co-variant in the variables.

The weight of an invariant is identical with the number of the bonds in the chemicograph of the analogous chemical substance, and the weight of the leading term (or basic differentiant) of a co-variant is the same as the number of bonds in the chemicograph of the analogous compound radical. Every invariant and covariant thus becomes expressible by a *graph* precisely identical with a Kekuléan diagram or chemicograph. But not every chemicograph is an algebraical one. I show that by an application of the algebraical law of reciprocity every algebraical graph of a given invariant will represent the constitution in terms of the roots of a quantic of a type reciprocal to that of the given invariant of an invariant belonging to that reciprocal type. I give a rule for the geometrical multiplication of graphs, that is, for constructing a *graph* to the product of in- or co-variants whose separate graphs are given. I have also ventured upon a hypothesis which, whilst in nowise interfering with existing chemicographical constructions, accounts for the seeming anomaly of the isolated existence as "monad molecules" of mercury, zinc, and arsenic—and gives a rational explanation of the "mutual saturation of bonds".

I have thus been led to see more clearly than ever I did before the existence of a common ground to the new mechanism, the new chemistry, and the new algebra. Underlying all these is the theory of pure colligation, which applies undistinguishably to the three great theories, all initiated within the last third of a century or thereabouts by Eisenstein, Kekulé, and Peaucellier.

Sylvester's paper in the *American Journal of Mathematics* (which he himself had founded) was entitled 'On an application of the new atomic theory to the graphical representation of the invariants and covariants of binary quantics' [14]. In it, his ideas are developed at length. A few sentences from one of the appendices are worth quoting:

> Chemistry has the same quickening and suggestive influence upon the algebraist as a visit to the Royal Academy, or the old masters may be supposed to have on a Browning or a Tennyson. Indeed it seems to me that an exact homology exists between painting and poetry on the one hand and modern chemistry and modern algebra on the other. In poetry and algebra we have the pure idea elaborated and expressed through the vehicle of language, in painting and chemistry the idea enveloped in matter, depending in part on manual processes and the resources of art for its due manifestation.

In spite of these enthusiastic sentiments, it is worth noting that Sylvester was somewhat nervous of the reception his paper might receive. In a letter to Simon Newcomb [15, p. 134], he wrote:

> I feel anxious as to how it will be received as it will be thought by many strained and over-fanciful. It is more a 'reverie' than a regular mathematical paper. I have however added some supplementary mathematical matter which will I hope serve to rescue the chemical portion from absolute

contempt. It may at the worst serve to suggest to chemists and Algebraists that they may have something to learn from each other.

A debate on the merits of the 'chemico-algebraic' theory ensued in the pages of the *American Journal*, and several eminent chemists and mathematicians contributed letters. One such letter was from Clifford [16], and to this letter Sylvester added many footnotes; the two men vied to ascribe to each other the credit for the discovery of the theory.

Unfortunately Sylvester's cherished theory was short-lived. What he had noticed was but a superficial analogy—the use of a similar notation— between two completely unrelated topics. Of course, the usefulness of the graphic notation in chemistry is unchallenged to this day, and in the theory of invariants it proved helpful for a while. Clifford himself continued to work on the subject [17], [18], until his researches were cut short by his untimely death, and various English mathematicians loyally produced several studies on 'Clifford's graphs' [19], [20]. But the whole cottage industry of invariant theory was soon taken over by the powerful methods of abstract algebra, heralded by D. Hilbert's proof [21] of the 'finite basis theorem'.

Although the chemico-algebraic theory was a flop, it was a major stimulus for the study of graphs as objects of interest in themselves. In this respect, a landmark in graph theory was the publication, in 1891, of the work of the Danish mathematician, Julius PETERSEN, who had frequently corresponded with Sylvester. We shall describe Petersen's achievements in Chapter 10.

Enumeration, from Cayley to Pólya

Cayley's method for enumerating hydrocarbons, explained in his 1875 address to the British Association, was cumbersome and impractical; perhaps this was one reason why it made little impact on the chemists of his own country. However, a shortened account of his method was also published in German [11], and drew an immediate response from H. Schiff [22], [23], who reported on an alternative method of enumeration. During the next thirty years, several chemists and mathematicians considered the problem, but their publications do not contain any significant new developments in the mathematical theory; a survey of some of this work may be found in [24].

It was not until the late 1920s that the problem was once more subjected to searching mathematical analysis. In fact, the first work which might have had important applications was published by J. H. Redfield in 1927 [25], but his paper was unfortunately completely overlooked for many years, and so had no influence on the development of the subject. However, a paper which did arouse some interest was published in 1929, under the title 'Isomerism and Configuration' [26]. Its authors were a chemist, A. C. Lunn, and a mathematician, J. K. Senior; they recognized that the terminology of the mathematical theory of permutation groups is appropriate to the problem of enumerating isomers, and in this respect, their paper foreshadowed the fundamental work of Pólya.

Before we come to Pólya's work, we should mention a series of papers by two chemists, C. M. Blair and H. R. Henze, which appeared from 1931 onwards (see, for example [24], [27]). They realized that it was too optimistic to hope for a formula giving the numbers of isomers in a series as a function of the carbon content, and so they concentrated instead on setting up recursion formulae for these numbers. This straightforward mathematical technique works quite well in many cases, and, in recent years, its scope has been considerably extended by the use of computers.

The enumerative theory due to George Pólya is a milestone, not only in graph theory and chemical enumeration, but in mathematics as a whole. Much of the work was developed in the period covered by this book, and we shall reproduce one of Pólya's earlier papers [4D] on the subject, which was published in 1935.

We introduce Pólya's paper by means of a simple example. Let us suppose that each of the vertices of a square is to be coloured with one of two colours, black and white. The total number of ways in which this can be done is $2^4 = 16$, but some of these colourings are indistinguishable if we take into account the symmetry of the square. For example, the two colourings shown in Fig. 4.5 may be transformed into one another by giving the square a half-turn about its centre. It turns out that there are just six non-equivalent colourings in this sense, and we have depicted these in Fig. 4.6.

In order to deal with enumerative problems of this kind in general, Pólya combined the classical method of generating functions with some basic results from the theory of permutation groups. His method can be outlined in the following way, using the notation of [4D] and referring to the example given above. We begin with a set of 'figures', each of which contains a number of 'objects'; in our example, a figure contains just one object—a vertex—and it may be coloured black or white, so that the

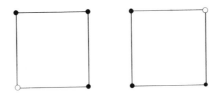

Fig. 4.5.

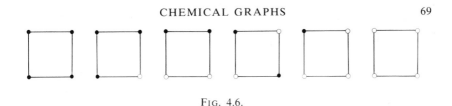

FIG. 4.6.

generating function for the set of figures is simply

$$f(x, y) = x + y.$$

(In general, the coefficient a_{ij} of the term $x^i y^j$ is the number of figures containing i black vertices and j white vertices; here $a_{10} = a_{01} = 1$, $a_{00} = a_{11} = 0$). We next introduce the idea of a 'configuration', which is formed by placing a figure at each of a given set of points; in our example, the points are the corners of a square. Two configurations are 'equivalent' if they are related by a symmetry of the set of points.

Our aim is to find the generating function for the set of non-equivalent configurations. That is, we require the function

$$\mathcal{F}(x, y) = \sum A_{ij} x^i y^j,$$

where A_{ij} denotes the number of non-equivalent configurations containing i black objects and j white ones. In order to explain the relationship between $\mathcal{F}$ and f, we must consider the group of permutations acting as symmetries of the given set of points. In our example, there are eight such permutations, and they are, in the usual cycle form:

$$(1)(2)(3)(4), \quad (1)(3)(24), \quad (2)(4)(13), \quad (13)(24),$$
$$(12)(34), \quad (14)(23), \quad (1234), \quad (1432).$$

(The numbers 1, 2, 3, 4, stand for the corners of the square in clockwise order.) In general, the group will contain n permutations in all, and the number of permutations with j_1 1-cycles, j_2 2-cycles, ..., and j_p p-cycles will be denoted by $H_{j_1 j_2 \ldots j_p}$. The *cycle index* of the group is then defined to be the expression

$$\frac{1}{n} \sum H_{j_1 j_2 \ldots j_p} s_1^{j_1} s_2^{j_2} \ldots s_p^{j_p},$$

where the sum is over all possible choices of the j_i for permutations in the group. For example, the cycle index for the group of the square is

$$\tfrac{1}{8}(s_1^4 + 2 s_1^2 s_2 + 3 s_2^2 + 2 s_4).$$

We can now state Pólya's famous result: *the generating function $\mathcal{F}(x, y)$ is obtained by substituting $f(x^r, y^r)$ for s_r in the cycle index.*

In the case of the square, this gives

$$\mathscr{F}(x, y) = \tfrac{1}{8}\{(x+y)^4 + 2(x+y)^2(x^2+y^2) + 3(x^2+y^2)^2 + 2(x^4+y^4)\}$$

$$= x^4 + x^3 y + 2x^2 y^2 + xy^3 + y^4,$$

from which we deduce (looking at the coefficients) that there are, for example, two configurations with two black vertices and two white vertices, and one configuration with three black vertices and one white vertex (see Fig. 4.6).

In [4D], the theory is developed for three colours instead of two, and there is an example in which only one colour is needed.

4D

G. PÓLYA
UN PROBLÈME COMBINATOIRE GÉNÉRAL SUR LES GROUPES DE PERMUTA-
TIONS ET LE CALCUL DU NOMBRE DES ISOMÈRES DES COMPOSÉS
ORGANIQUES

[A general combinatorial problem on groups of permutations and the calculation
of the number of isomers of organic compounds]

Comptes Rendus Hebdomadaires des Séances de l'Académie des Sciences (*Paris*)
201 (1935), 1167–1169.

I shall state my problem in concrete language and in a slightly specialised form.

1. We are given figures Φ, Φ^*, Φ^{**}, ..., each different from one another. These figures contain three sorts of objects, some *red*, some *blue* and some *white*. The number of figures containing k red, l blue and m white is a_{klm}. We put

$$\sum a_{klm} x^k y^l z^m = f(x, y, z),$$
$$f(x, y, z) = f_1, \quad f(x^2, y^2, z^2) = f_2, \quad f(x^3, y^3, z^3) = f_3, \ldots.$$

2. We are given p points fixed in space and a group $\mathcal{H}$ of order h permuting these points. Let $H_{j_1 j_2 \ldots j_p}$ be the number of permutations of $\mathcal{H}$ fixing j_1 points, and effecting j_2 transpositions, j_3 cycles of order 3, j_4 cycles of order 4, etc. These permutations are said to be of *type* $[j_1, j_2, \ldots, j_p]$; naturally, $1j_1 + 2j_2 + \ldots + pj_p = p$.

3. By placing p figures from amongst Φ, Φ^*, Φ^{**}, ... at the p points, we obtain a *configuration* $(\Phi_1, \Phi_2, \ldots, \Phi_p)$; the figures may be repeated, that is to say the same figure may appear at several points in the same configuration. Two configurations $(\Phi_1, \Phi_2, \ldots, \Phi_p)$ and $(\Phi_1', \Phi_2', \ldots, \Phi_p')$ are identical if $\Phi_1 = \Phi_1'$, $\Phi_2 = \Phi_2', \ldots, \Phi_p = \Phi_p'$, and they are called *equivalent* mod $\mathcal{H}$ if there exists a permutation

(1)
$$S_i = \begin{pmatrix} 1 & 2 \ldots p \\ i_1 & i_2 \ldots i_p \end{pmatrix}$$

of the group $\mathcal{H}$ such that $\Phi_{i_1} = \Phi_1'$, $\Phi_{i_2} = \Phi_2', \ldots, \Phi_{i_p} = \Phi_p'$.

4. We seek *the number* A_{klm} *of non-equivalent configurations* mod $\mathcal{H}$ *which contain* k *red,* l *blue* and *m* *white objects.* We are going to find the generating function

$$\sum A_{klm} x^k y^l z^m = \mathcal{F}(x, y, z).$$

5. We say that the configuration $(\Phi_1, \Phi_2, \ldots, \Phi_p)$ *admits the permutation* (1) if $\Phi_1 = \Phi_{i_1}, \Phi_2 = \Phi_{i_2}, \ldots, \Phi_p = \Phi_{i_p}$). Let $A_{klm}(S_i)$ be the number of configurations admitting S_i, and containing k red, l blue and m white objects. If the type of S_i is $[j_1, j_2, \ldots, j_p]$ it is easily seen that

(2)
$$\sum A_{klm}(S_i) x^k y^l z^m = f_1^{j_1} f_2^{j_2} \ldots f_p^{j_p};$$

this is just the classical method of Euler.

6. The permutations of $\mathcal{H}$ admitted by a fixed configuration C form a subgroup $\mathcal{G}$ of order g. The number of configurations different from one another, but equivalent to C, is h/g, and each of these configurations admits a subgroup, whose order is g. Each of these h/g configurations contains the same numbers k, l, m of red, blue and white objects, and each is counted in g terms of the sum

$$A_{klm}(S_1) + A_{klm}(S_2) + \ldots + A_{klm}(S_h).$$

Thus in this sum, the family of configurations equivalent to C (mod $\mathcal{H}$) is counted $g h/g = h$ times, that is to say, we have

(3)
$$A_{klm}(S_1) + A_{klm}(S_2) + \ldots + A_{klm}(S_h) = h A_{klm}.$$

We obtain from (2) and (3),

(4)
$$\mathcal{F}(x, y, z) = \frac{1}{h} \sum H_{j_1 j_2 \ldots j_p} f_1^{j_1} f_2^{j_2} \cdots f_p^{j_p},$$

the sum ranging over the types $[j_1, j_2, \ldots, j_p]$.

7. The formula (4) has a large number of applications in the study of symmetries, in particular in the calculation of the number of isomers, of which I shall speak elsewhere. Here I restrict myself to a single example. The *figures* are the radicals $C_n H_{2n+1}$, $n = 0$ included, the objects are the atoms C (there is only one type of object), and a_n is the number of isomeric alcohols $C_n H_{2n+1} OH$ (without taking account of stereoisomerism). Thus we have

$$f(x) = 1 + x + x^2 + 2x^3 + 4x^4 + 8x^5 + 17x^6 + \ldots.$$

I take $p = 6$ points, vertices of a regular hexagon, and $\mathcal{H}$ to be the group of $h = 12$ rotations of the hexagon. The coefficient of x^n in the expansion of

$$\frac{1}{12}[f^6 + 4f_2^3 + 3f_1^2 f_2^2 + 2f_3^2 + 2f_6]$$

$$= 1 + x + 4x^2 + 8x^3 + 22x^4 + 51x^5 + 136x^6 + \ldots$$

is then the number of derived isomers $C_{6+n} H_{6+2n}$ of benzene.

Pólya wrote several other papers on this subject. He published an account of the applications to isomer enumeration in *Zeitschrift für Kristallographie* [28], and in 1937 a long paper of over one hundred pages appeared in *Acta Mathematica* [29]. In the latter paper he discussed many mathematical aspects of the theory, including its application to the enumeration of graphs.

We should not conclude this chapter without some further reference to the work of Redfield. His 1927 paper (the only one he ever wrote) contained many of the ideas later rediscovered independently by Pólya. In particular, towards the end of his paper he dealt with the problem of enumerating what he called 'symmetrical aliorelative dyadic relation-numbers of a field of n elements'; in our terms, these objects are just simple graphs with n vertices. Perhaps this language difficulty may explain why so many English-speaking mathematicians are so much more familiar with the results of Pólya's German paper than with those of Redfield's English one.

References

1. KEKULÉ, F. A. [Antwort des Jubilars August Kekulé]. *Ber. Deut. Chem. Ges.* **23** (1890), 1306.
2. RUSSELL, C. A. *The history of valency.* Leicester University Press, 1971.
3. WHEELER, T. S. and PARTINGTON, J. R. *The life and work of William Higgins, chemist.* Pergamon Press, Oxford, 1960.
4. COUPER, A. S. Sur une nouvelle théorie chimique. *Phil. Mag.* (4) **16** (1858), 104–116.
5. LOSCHMIDT, J. *Chemische Studien.* Wien, 1861.
6. KEKULÉ, F. A. *Lehrbuch der organischen Chemie*, Vol. 1. Erlangen, 1861.
7. CRUM BROWN, A. On the theory of isomeric compounds. *Trans. Roy. Soc. Edinb.* **23** (1864), 707–719.
8. CAYLEY, A. On the theory of the analytical forms called trees—part II. *Phil. Mag.* (4) **18** (1859), 374–378 = *Math. Papers*, Vol. 4, 112–115.
9. SCHORLEMMER, C. *A manual of the chemistry of the carbon compounds.* London, 1874.
10. CAYLEY, A. On the analytical forms called trees, with application to the theory of chemical combinations. *Rep. Brit. Assoc. Advance. Sci.* **45** (1875), 257–305 = *Math. Papers*, Vol. 9, 427–460.
11. CAYLEY, A. Über die analytischen Figuren, welche in der Mathematik Bäume genannt werden und ihre Anwendung auf die Theorie chemischer Verbindungen. *Ber. Deut. Chem. Ges.* **8** (1875), 1056–1059.
12. CAYLEY, A. On the number of the univalent radicals $C_n H_{2n+1}$. *Phil. Mag.* (5) **3** (1877), 34–35 = *Math. Papers*, Vol. 9, 544–545.
13. [Report of the British Association Meeting]. *Athenaeum* No. 2498 (11 Sept. 1875), 341.
14. SYLVESTER, J. J. On an application of the new atomic theory to the graphical representation of the invariants and covariants of binary quantics,—with three appendices. *Amer. J. Math.* **1** (1878), 64–125 = *Math. Papers*, Vol. 3, 148–206.
15. ARCHIBALD, R. C. Unpublished letters of James Joseph Sylvester... *Osiris* **1** (1936), 85–154.
16. CLIFFORD, W. K. Extract of a letter to Mr. Sylvester... *Amer. J. Math.* **1** (1878), 126–128 = *Math. Papers*, 255–257.
17. CLIFFORD, W. K. Note on quantics. *Proc. London Math. Soc.* **10** (1878–9), 124–129 = *Math. Papers*, 258–265.
18. CLIFFORD, W. K. Binary forms of alternate variables. *Proc. London Math. Soc.* **10** (1878–9), 214–221 = *Math. Papers* 277–286.
19. SPOTTISWOODE, W. On Clifford's graphs. *Proc. London Math. Soc.* **10** (1878–9), 204–214.

20. KEMPE, A. B. On an application of Clifford's graphs to ordinary binary quantics. *Proc. London Math. Soc.* **17** (1887), 107–121.

21. HILBERT, D. Zur Theorie der algebraischen Gebilde I. *Nachr. K. Ges. Wiss. Göttingen*, (1888), 450–457 = *Ges. Abh.*, Vol. 2, 176-183.

22. SCHIFF, H. [Correspondenzen] aus Florenz. *Ber. Deut. Chem. Ges.* **8** (1875), 1360–1361.

23. SCHIFF, H. Zur Statistik chemischer Verbindungen. *Ber. Deut. Chem. Ges.* **8** (1875), 1542–1547.

24. HENZE, H. R. and BLAIR, C. M. The number of isomeric hydrocarbons of the methane series. *J. Amer. Chem. Soc.* **53** (1931), 3077–3085.

25. REDFIELD, J. H. The theory of group-reduced distributions. *Amer. J. Math.* **49** (1927), 433–455.

26. LUNN, A. C. and SENIOR, J. K. Isomerism and configuration. *J. Phys. Chem.* **33** (1929), 1027–1079.

27. HENZE, H. R. and BLAIR, C. M. The number of structural isomers of the more important types of aliphatic compounds, *J. Amer. Chem. Soc.* **56** (1934), 157.

28. PÓLYA, G. Algebraische Berechnung der Anzahl der Isomeren einiger organischer Verbindungen. *Z. Kristal.* (A) **93** (1936), 415–443.

29. PÓLYA, G. Kombinatorische Anzahlbestimmungen fur Gruppen, Graphen und chemische Verbindungen. *Acta. Math.* **68** (1937), 145–254.

5

Euler's Polyhedral Formula

A.-L. CAUCHY (1789–1857)

IN our account of the work of Kirkman and Hamilton (Chapter 2) we explained how a solid figure bounded by plane faces, called a polyhedron, can be represented by a graph. In this chapter we shall trace the study of polyhedra and their graphs, taking as our theme one important formula and its generalizations. The simplest version of this formula relates the numbers of vertices (N_v), edges (N_e), and faces (N_f) of a polyhedron; under certain quite general conditions it asserts that

$$N_v - N_e + N_f = 2.$$

For example, for the pentagonal pyramid (Fig. 5.1) the formula yields $6 - 10 + 6 = 2$, and for the regular dodecahedron (Fig. 5.2) we have $12 - 30 + 20 = 2$.

The history of polyhedra

Although the study of polyhedra can be traced back at least four thousand years, to the pyramids of Ancient Egypt, we shall begin our story some two thousand years later, in Greece. The Greeks were especially interested in the mathematical properties of 'regular' polyhedra (those in which all the faces are congruent regular polygons), and they found just five of these solids; a full account of them was written by

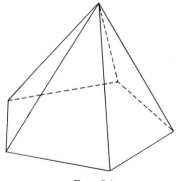

FIG. 5.1.

Theaetetus, who died in 369 B.C. What is more, it has been argued that the *Elements* of Euclid (written about 320 B.C.) were primarily intended as an introduction to the study of these five solids. But surprisingly, there is no evidence that the Greeks knew the simple formula relating the numbers of vertices, edges, and faces.

There are two possible reasons for this. One possibility is that they did discover the formula, and that it was lost in subsequent, more barbarous, times. We know that many works of Greek scholarship did not survive; in particular, some of the writings of Archimedes (287–212 B.C.) were destroyed, probably in the fire of Alexandria (47 B.C). Archimedes was one of the greatest of the Greek thinkers, and he had made an extensive study of polyhedra. The second possibility is that the Greeks missed the formula because their geometry was entirely concerned with measurement—lengths, angles, areas, and volumes. The formula is not about such metrical notions; it is, in fact, a topological result.

In comparatively recent times (around 1640 A.D.), René Descartes, the founder of analytic geometry, also missed the formula. He obtained an expression for the sum of the angles of all the faces of a polyhedron, from

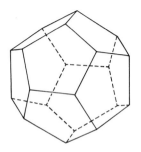

FIG. 5.2.

which the required formula can be deduced. But he did not make the deduction, presumably because he too was accustomed to thinking in terms of measurement. We know of Descartes' work because Leibniz made a copy of it, but this was not discovered until 1859. The text, together with the story of its discovery, translation, and publication, may be found in the collected works of Descartes [1, Vol. 10, pp. 257–277].

By the time that Descartes' work came to light, the formula he had missed so narrowly was well known, for it had been discovered and published by Euler. For many years Euler corresponded with a fellow mathematician, C. Goldbach, and in 1750 he wrote him a letter [5A] saying that he had found the formula, but had not been able to prove it. The complete letter, written partly in German and partly in Latin, is available in the Euler–Goldbach correspondence [2]. We have omitted the first part (dealing with a dispute between Goldbach and a bookseller) and the conclusion, in which Euler gave a formula for the volume of a tetrahedron in terms of its edges.

5A

L. EULER
[Letter to Christian Goldbach]

Berlin, November 1750.

*   *   *   *   *

Recently it occurred to me to determine the general properties of solids bounded by plane faces, because there is no doubt that general theorems should be found for them, just as for plane rectilinear figures, whose properties are: (1) that in every plane figure the number of sides is equal to the number of angles, and (2) that the sum of all the angles is equal to twice as many right angles as there are sides, less four. Whereas for plane figures only sides and angles need to be considered, for the case of solids more parts must be taken into account, namely

  I. the faces, whose number $= H$;
 II. the solid angles, whose number $= S$;
III. the joints where two faces come together side to side, which, for lack of an accepted word I call 'edges', whose number $= A$;
 IV. the sides of all the faces, the number of which all added together $= L$;
  V. the plane angles of all faces, the total number of which $= P$.

1. Concerning these five quantities, it is clear first that $P = L$, since in each face the number of angles $=$ the number of sides.
2. Also, A always $= \frac{1}{2}L$, or $A = \frac{1}{2}P$, because two sides of different faces must come together to form one edge.
3. Therefore the number of sides, or of plane angles, of all the faces enclosing the solid is always even.

4. Either $L = 3H$ or $L > 3H$
5. Either $P = 3S$ or $P > 3S$ } and $P = L$.

These are obvious, since no face can have less than 3 sides, and no solid angle less than 3 plane angles. But I cannot yet give an entirely satisfactory proof of the following propositions:

6. In every solid enclosed by plane faces the aggregate of the number of faces and the number of solid angles exceeds by two the number of edges, or $H + S = A + 2$, or $H + S = \frac{1}{2}L + 2 = \frac{1}{2}P + 2$.

7. It is impossible that $A + 6 > 3H$ or $A + 6 > 3S$.

8. It is impossible that $H + 4 > 2S$ or $S + 4 > 2H$.

9. No solid can be formed all of whose faces have 6 or more sides, nor all of whose solid angles are joined from six or more plane angles.

10. The sum of all plane angles which occur in the solid is equal to as many right angles as there are units in $4A - 4H$.

11. The sum of all plane angles is equal to four times as many right angles as there are solid angles, less eight, that is $= 4S - 8$ right angles.

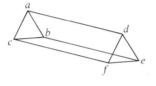

FIG. 5.3.

For an example consider the triangular prism [see Fig. 5.3], where:
1. the number of faces, $H = 5$;
2. the number of solid angles, $S = 6$;
3. the number of edges (ab, ac, bc, ad, be, cf, de, df, ef), $A = 9$;
4. the number of sides and plane angles, $L = P = 18$. For the body is bounded by two triangles and three quadrilaterals, so that $L = P = 2.3 + 3.4 = 18$.

Now, in agreement with Theorem 6, $H + S(11) = A + 2(11)$. Furthermore the sum of all plane angles (from the two triangles $= 4$ right angles, and from the three quadrilaterals $= 12$ right angles) is $= 16$ right angles, $= 4(A - H) = 4S - 8$ right angles.

I find it surprising that these general results in solid geometry have not previously been noticed by anyone, so far as I am aware; and furthermore, that the important ones, Theorems 6 and 11, are so difficult that I have not yet been able to prove them in a satisfactory way.

*   *   *   *   *

In 1752 Euler wrote two papers on the polyhedral formula. In the first one [3], he verified it for several families of solids, but admitted that he had still been unable to find a general proof. In the second one [4], he set forth a proof by dissection; it consists of slicing out tetrahedral pieces of the solid in such a way that the number $N_v - N_e + N_f$ is unchanged, until eventually all that remains is a single tetrahedron, for which $N_v - N_e + N_f$ $= 4 - 6 + 4 = 2$. It is an ingenious method—and an apt one, since it uses no metrical concepts. However, it is not at all obvious that this slicing procedure can always be carried out, and it may give rise to 'degenerate' polyhedra for which the meaning of the formula is ambiguous.

One point which was overlooked by Euler, and some other early writers on the subject of polyhedra, is the matter of convexity. A solid body S is said to be *convex* if, for any two points X and Y of S, the entire line segment XY lies in S. Although this property holds for all the polyhedra considered by the Greeks, it is by no means universal. In 1619 the astronomer Johannes Kepler [5] described a non-convex polyhedron which he called the *stella octangula* (Fig. 5.4); for further details of this and many other fascinating non-convex polyhedra the interested reader is referred to the book by Wenninger [6]. We do not need to discuss convexity in detail, but it should be remembered that this property cannot be taken for granted in the theory of polyhedra.

An entirely different proof of Euler's formula was given by A. M. Legendre in 1794 [7, p. 228] using metrical properties of spherical polygons. In fact, his ideas were quite similar to those used by Descartes many years earlier, but Legendre had the advantage of knowing what he wanted to prove, and so he pursued the arguments to their logical conclusion.

Planar graphs and maps

We have already referred at some length to a paper [8] written by Poinsot in 1809, in which he discussed some diagram-tracing puzzles (see Chapter 1). One of the major achievements of that paper was the description of four non-convex regular polyhedra, in addition to the five convex ones known to the Greeks; all nine polyhedra have either 4, 6, 8, 12, or 20 faces, and Poinsot asked if these were the only possibilities.

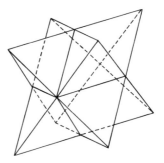

FIG. 5.4.

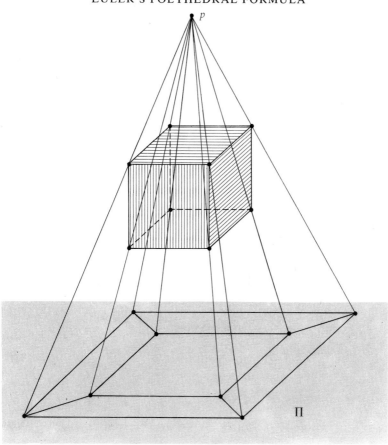

FIG. 5.5.

Soon afterwards, a brilliant young Frenchman, Augustin-Louis CAUCHY, proved that this is indeed so. In the course of his proof he investigated certain new aspects of Euler's formula, one of which is particularly interesting to us, since it involves the formulation of the result as a theorem about graphs drawn in the plane. This idea is central in later chapters of this book, and so we shall devote some time to it here.

In Fig. 5.5, we depict the process of projecting a cube from a point p on to a plane Π. The vertices and edges of the cube are projected into a plane diagram of the graph associated with the cube, whereas the faces correspond to the regions into which the plane is divided by the lines of the diagram. We remark that one face (the one nearest the point p) is lost—it corresponds to the outside region of the diagram. It is not hard to see that, for any convex polyhedron, the point p and the plane Π can be

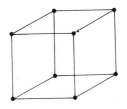

FIG. 5.6.

chosen so that the projected diagram is a faithful representation of the graph of the polyhedron. By this we mean that the lines representing the edges intersect only at the points representing the vertices, so that no extra crossings occur. In this sense, the diagram in Fig. 5.6 is not a faithful representation of the cube graph.

We say that a graph is **planar** if it has a faithful representation by means of a plane diagram. The process of projection shows that the graph of a convex polyhedron is planar, although (as we shall see in Chapter 8) there are many abstract graphs which are not planar.

It is also possible to think of the plane diagram of a graph as a geographical map: the regions are the 'countries' and the lines are their 'frontiers'. Of course, the same kind of map may be drawn on other surfaces besides the plane; for example, if we take a point inside a convex polyhedron and project from it onto a sphere surrounding the polyhedron, we obtain a 'map' on the sphere. The fact that the edges are no longer represented by straight lines is immaterial, for it is necessary only that the representation is faithful, and that no extra crossings occur. In general we shall define a **map** to be a representation of a graph on any surface, with the property that the lines intersect only at their assigned end points.

With this terminology, Cauchy's formulation of Euler's result is a theorem about planar graphs, or equivalently, about maps on the plane. Our extract [5B] begins with a generalization of Euler's formula. Cauchy considered the effect of allowing extra vertices and edges inside a polyhedron so that it is, in fact, decomposed into a number P of separate polyhedra. For this situation he stated that

$$N_v - N_e + N_f = P + 1,$$

from which the Euler formula is obtained by putting $P = 1$. He continued by discussing the effect of flattening the polyhedron so that it becomes a plane network of polygons, and he explained how this could be regarded as the case $P = 0$ of his general formula. He then gave a direct proof of the relation

$$N_v - N_e + N_f = 1$$

for this case. As the reader will see, his proof really does belong to graph theory, for it uses a process of 'triangulation', rather than any metrical ideas. So this procedure, together with the process of projection, can be used to give a topological proof of the original polyhedral formula.

5B

A.-L. CAUCHY

RECHERCHES SUR LES POLYÈDRES—PREMIER MÉMOIRE

[Researches on polyhedra—first memoir]

Journal de l'École Polytechnique **9** (Cah. 16) (1813), 68–86.

*   *   *   *   *

SECOND PART

Euler was the first to determine, in the Petersburg *Memoirs* for 1758, the relation which holds between the different elements forming the surface of a polyhedron; and Legendre, in his *Éléments de Géométrie*, has proved Euler's theorem in a much simpler manner by considering spherical polygons. Having been led by some research to a new proof of the theorem, I have arrived at a more general theorem than that of Euler, which is stated as follows:

THEOREM.—*If a polyhedron is decomposed into as many others as we choose, by taking at will new vertices in the interior, and if the number of new polyhedra so formed is denoted by* P, *the total number of vertices including those of the original polyhedron by* S, *the total number of faces by* F, *and the total number of edges by* A, *then*

$$(1) \qquad\qquad S+F=A+P+1,$$

that is, the sum of the number of vertices and that of the faces exceeds by one unit the sum of the number of edges and that of the polyhedra.

It is easy to see that Euler's theorem is a special case of the preceding theorem; since, if we suppose that all the polyhedra reduce to a single one, then we have

$$P=1,$$

and equation (1) becomes

$$(2) \qquad\qquad S+F=A+2.$$

Furthermore, we may deduce from equation (1) a second theorem relating to plane geometry; for, supposing that all the polyhedra are reduced to a single one, let us destroy this last one by taking one of its faces as base and projecting onto it all the other vertices without changing their number; then we obtain a plane figure composed of several polygons enclosed in a given contour. Let F be the number of these polygons, S the number of their vertices, and A that of their sides; we shall obtain the relationship between these three numbers by putting $P=0$ in the general formula, so that we have

$$(3) \qquad\qquad S+F=A+1.$$

From this we conclude that the sum of the number of polygons and the number of vertices exceeds by one unit the number of straight lines which form the sides of these polygons. This last theorem is the equivalent in plane geometry of the general theorem in the geometry of polyhedra.

We could prove the general theorem contained in equation (1) immediately, and deduce from it as corollaries the two other theorems. But in order to explain the spirit of the proof more clearly, we shall begin by proving, in an analogous manner, the last theorem contained in equation (3).

It is easy to apply the theorem in certain particular cases.

Let us suppose, for example, that the contour is given as the perimeter of a triangle, that we take a point in the interior of it, and that we construct three straight lines from this point to the three vertices, thereby forming three triangles inside the given contour. These three triangles furnish four vertices, and the number of straight lines forming their sides is equal to 6; now 6, increased by one, gives the same as $4+3$, which verifies the theorem.

Suppose, as a second case, that the contour is given as a quadrilateral, that we take a point in its interior, and that we construct four straight lines from this point to the four vertices, thereby forming four triangles in the given contour. These four triangles furnish five vertices and eight sides, so

$$8+1=4+5,$$

which verifies the theorem.

Finally, suppose that the contour is given as a polygon of n sides, and that we take a point in its interior which we join to the n vertices of the polygon by means of straight lines. The n triangles which are formed by this means furnish a number of vertices equal to $n+1$, and a number of sides equal to $2n$; then $2n$ increased by one is equal to the sum of n and $n+1$, which verifies the theorem.

Now let us pass to the general case, and suppose that a number F of polygons is enclosed in a given contour. Let S be the number of vertices of these polygons and A be the number of straight lines which form their sides. Decompose each polygon into triangles, by joining one of its vertices to non-adjacent vertices by diagonals. If n is the number of diagonals drawn in the different polygons, $F+n$ will be the number of triangles resulting from the decomposition of the polygons, and $A+n$ will be the number of sides of these triangles. The number of their vertices will be the same as for the vertices of the polygons, that is, S. Now let us suppose that we remove the triangles in turn, in such a way as to leave just one at the end, beginning with those which adjoin the exterior contour and subsequently removing only those which, as a result of previous reductions, have one or two sides belonging to that contour. Let h' be the number of triangles which have one side in common with the exterior contour at the time they are removed, and h'' the number of triangles which have two sides in common with that contour. The destruction of each triangle will entail, in the first case, the loss of one side; and in the second case, the loss of two sides and one vertex. It follows that, when all the triangles except one have been destroyed, the number of triangles destroyed will be

$$h'+h'',$$

the number of sides removed will be

$$h'+2h'',$$

and the number of vertices removed will be

$$h''.$$

The number of triangles remaining will therefore be

$$F + n - (h' + h'') = 1,$$

the number of sides remaining will be

$$A + n - (h' + 2h'') = 3,$$

and the number of vertices remaining will be

$$S - h'' = 3.$$

If we add the first equation to the third and subtract the second, we shall have

$$S + F - A = 1,$$

or

(3) $$S + F = A + 1,$$

which was to be proved.

*   *   *   *   *

Cauchy's memoir also contained a general proof of his extension of Euler's formula, but we turn now to other generalizations, more relevant to graph theory.

Generalizations of Euler's formula

At about the same time as Cauchy was writing his memoir, a systematic study of exceptions to Euler's formula was being made by the Professor of Mathematics in Geneva, Simon-Antoine-Jean LHUILIER. Lhuilier found three types of polyhedra for which the formula fails, and he developed the generalizations needed to deal with them. He published a paper [9] on this topic in 1811, and sent a long memoir containing his results to J. D. Gergonne, a French mathematician who had founded his own journal; however, the memoir was so long that Gergonne could not publish all of it. Instead he published his own account of Lhuilier's memoir into which he incorporated some of his own ideas. The next extract [5C] is taken from this paper.

The first part of our extract is concerned with the application of Euler's formula to the proof of the fact that there are only five regular convex polyhedra. It is probable that this work was common knowledge among mathematicians at that time, but the account is notable for three reasons. First, it points out that the derivation makes no appeal to congruence or other metrical notions. Secondly, the concept of duality is anticipated by the remark that regular polyhedra occur in reciprocal pairs; and thirdly, the regular subdivisions (tessellations) of the plane are treated as limiting cases of regular polyhedra in which the number of faces has become infinite.

The second part of our extract deals with exceptions to Euler's formula; these exceptions lead on to important topological considerations about surfaces, which will be discussed after the paper.

5C

S.-A.-J. LHUILIER

MÉMOIRE SUR LA POLYÉDROMÉTRIE

[Memoir on polyhedrometry]

Annales de Mathématiques **3** (1812–3), 169–189.

$$* \quad * \quad * \quad * \quad *$$

8. If one requires that in a polyhedron all the faces have the same number f of sides and all the solid angles the same number s of edges, then in order to determine A, F, and S we have the three equations:

$$fF = 2A, \qquad sS = 2A, \qquad S + F = A + 2.$$

If f and s, F and S, are interchanged these equations remain the same, and we conclude that the polyhedra of this kind are reciprocal in pairs; thus, in the two members of a pair, the number of edges is the same, and the number of vertices of each is equal to the number of faces of the other; this allows one to inscribe or circumscribe one in the other.

From these equations, we obtain

$$A = \frac{2fs}{2(f+s) - fs}, \qquad F = \frac{4s}{2(f+s) - fs}, \qquad S = \frac{4f}{2(f+s) - fs}.$$

The requirement that f, s, F, S, A be positive integers greater than 2, restricts the solutions of the equations as follows:

$$f = 3, \quad 3, \quad 4, \quad 3, \quad 5, \quad 3, \quad 6, \quad 4,$$
$$s = 3, \quad 4, \quad 3, \quad 5, \quad 3, \quad 6, \quad 3, \quad 4,$$
$$F = 4, \quad 8, \quad 6, \quad 20, \quad 12, \quad \infty, \quad \infty, \quad \infty,$$
$$S = 4, \quad 6, \quad 8, \quad 12, \quad 20, \quad \infty, \quad \infty, \quad \infty,$$
$$A = 6, \quad 12, \quad 12, \quad 30, \quad 30, \quad \infty, \quad \infty, \quad \infty.$$

We deduce from this not only that there are just five regular solids, but that there can be only five kinds of polyhedra, regular or not, with all faces having the same number of sides, and all solid angles the same number of edges.

We see also that there are just three different ways in which the sphere may be considered as a regular polyhedron with an infinite number of infinitely small faces; these faces may be either triangles joined six by six, or hexagons joined three by three, or finally, squares joined four by four.

We see also that a plane can be exactly covered with polygons of the same kind, assembled with the same number around each vertex, in just three different ways; with triangles assembled six by six, with squares assembled four by four, and with hexagons assembled three by three.

We see finally that the regular polyhedra occur in pairs, the tetrahedron with itself, the hexahedron with the octahedron, the dodecahedron with the icosahedron, the sphere covered by hexagons with the sphere covered by triangles, and the sphere covered by squares with itself.

$$*\quad *\quad *\quad *\quad *$$

As I have already stated, in the second part of his memoir Lhuilier concerns himself with the diverse exceptions to which *Euler's theorem* is subject. These exceptions are of three kinds. I shall present them successively.

11. The first kind of exception occurs when the polyhedron has an interior cavity; that is, if it has two separate surfaces and one of them entirely encloses the other.

In this case, let f, s, a be the numbers of faces, vertices, and edges of the exterior surface; and let f', s', a' be the analogous numbers for the interior surface; from the previous results we have

$$f+s = a+2, \qquad f'+s' = a'+2;$$

thus

$$(f+f')+(s+s') = (a+a')+4;$$

but, denoting by F the total number of faces of the polyhedron, by S the total number of vertices, and by A the total number of edges, we clearly have

$$f+f' = F,\ s+s' = S,\ a+a' = A;$$

it follows that

$$F+S = A+4;$$

that is to say, *in such a polyhedron, the number of faces augmented by the number of vertices, exceeds by four units the number of edges.*

In general, a solid may have n closed polyhedral surfaces exterior to one another, and a closed polyhedral surface enclosing them all; then, with the same notation as above, we have

$$F+S = A+2(n+1).$$

$$*\quad *\quad *\quad *\quad *$$

12. The second kind of exception occurs when the polyhedron is *ring-shaped;* that is if, although it has but a single surface, there is an opening passing right through it.

Let us imagine that we cut such a ring by a plane section which, by reason of the two faces separated by the cut, reduces it to the class of ordinary polyhedra; then let the number of faces be denoted by F', the number of vertices by S', and the number of edges by A'; as above, we have

$$F'+S' = A'+2.$$

Let n be the number of sides of the two faces of the cut, and let us imagine that these two faces are welded together in order to reconstruct the original polyhedron. Then denoting by S, F, A the quantities corresponding to those denoted by S' F', A' in the cut polyhedron, we find, by reasoning similar to that given above, that

$$F + S - A = F' + S' - A' - 2 = 2 - 2 = 0,$$

or

$$F + S = A;$$

that is, *in such a polyhedron, the number of faces augmented by the number of vertices is precisely equal to the number of edges.*

In general, a polyhedron bounded by a single surface may be pierced by an arbitrary number of distinct openings. If n denotes the number of these tunnels, we shall have

$$F + S = A - 2(n - 1).$$

* * * * *

The reasoning 'similar to that given above', referred to in the penultimate paragraph, is just the observation that $F' = F - 2$, $S' = S - v$, $A' = A - v$, where v denotes the number of sides of the cut face.

Lhuilier's third exception applies to polyhedra which have indentations in their faces, so that a face may appear like the first diagram in Fig. 5.7, rather than the second. In modern terminology a face of the second type is said to be *simply-connected*, whereas a face of the first type is not: Lhuilier remarked that faces which are not simply-connected occur naturally in certain crystals—Calcareous spar, for instance. But undoubtedly the most important of his three exceptions is the second one, and we shall now discuss it in some detail.

It is helpful to think of the formulae of Euler, Cauchy, and Lhuilier as statements about maps on various surfaces. We shall suppose that the surface is connected (that is, all in one piece, unlike Lhuilier's first

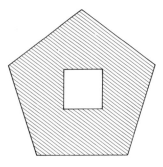
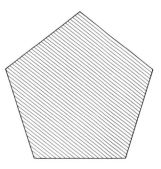

FIG. 5.7.

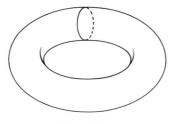

FIG. 5.8.

exceptions), and that every face is simply-connected (unlike his third exceptions). Then the formulae assert that the *Euler characteristic*,

$$\eta = N_v - N_e + N_f,$$

depends only on the surface on which the map is drawn, and not on the map itself. For example, the ordinary convex polyhedra of Euler correspond to maps on a sphere, for which $\eta = 2$, whereas the plane networks of Cauchy correspond to maps on a plane disc, for which $\eta = 1$.

The second class of exceptions considered by Lhuilier corresponds to maps on surfaces having one additional property besides those stated earlier: the surfaces are required to be *closed*, in the sense that they have no boundary curves and are of finite extent. So surfaces such as the whole plane, disc, or hemisphere, are not allowed. The simplest allowable surface is the sphere, and the next simplest is the ring-shaped surface referred to in [5C] and illustrated in Fig. 5.8. It is called the *torus*, or anchor ring. If we take a cube and bore a square tunnel joining two parallel faces, and add extra edges so that all of the faces are simply-connected, then we have an example of a polyhedron whose surface has the form of a torus (Fig. 5.9). The reader may care to check that $\eta = 0$ for this example, in agreement with Lhuilier's result.

In fact, infinitely many allowable surfaces may be constructed in a similar way: take a sphere and cut out two circular pieces from its surface; then take a cylindrical 'handle' and glue its ends to the sides of the holes

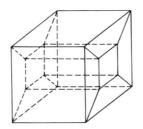

FIG. 5.9.

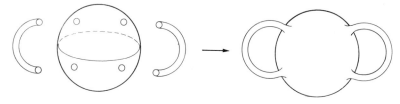

FIG. 5.10.

in the sphere; repeat this process g times, so that a sphere with g handles is obtained. The construction for $g = 2$ is shown in Fig. 5.10. The torus is just the case $g = 1$ and, of course, the sphere itself is the case $g = 0$. Our construction is equivalent to Lhuilier's procedure of boring g tunnels through a spherical polyhedron, and the result—a sphere with g handles—is usually known as a closed surface of *genus* g.

Nowadays we regard the last equation of [5C] as a formula for the Euler characteristic of a closed surface of genus g; that is, we assert that

$$\eta = 2 - 2g$$

for maps which can be drawn on such a surface. This formula was the starting point for an extensive investigation (1861) by Listing, which was to be influential in the separate development of topology as a branch of modern mathematics. His work, which he called *Der Census räumliche Complexe* (The census of spatial complexes) [10], continued the studies begun in the *Topologie*. He defined certain objects, called 'complexes' because they were built up from simpler pieces, and investigated their topological properties, especially the question of how such properties affect the generalizations of Euler's formula. The properties were given names such as 'periphraxis' and 'cyclosis', borrowed from the biological sciences. Listing's ideas were taken up by other mathematicians, and in particular, they were used by Poincaré in his fundamental researches on algebraic topology at the turn of the century. What Poincaré possessed, and Listing lacked, was the mathematical apparatus needed to formulate topological ideas within the framework of algebra. Poincaré's work will be discussed in more detail in Chapter 8.

Lhuilier had noticed all the peculiarities which a surface may exhibit, with one important omission; he overlooked the possibility that a surface can be 'one-sided'. This phenomenon of one-sidedness, exemplified by such surfaces as the Möbius strip and the Klein bottle, was not discovered until around 1858, as we shall see in Chapter 7.

References

1. ADAM, C. and TANNERY, P. *Oeuvres de Descartes*. Cerf, Paris, 1897–1913.
2. JUŠKEVIČ, A. P. and WINTER, E. *Leonhard Euler und Christian Goldbach: Briefwechsel 1729–1764*. Akademie–Verlag, Berlin, 1965.

3. EULER, L. Elementa doctrinae solidorum. *Novi Comm. Acad. Sci. Imp. Petropol.* **4** (1752–3, published 1758), 109–140 = *Opera Omnia* (1), Vol 26, 72–93.

4. EULER, L. Demonstratio nonnullarum insignium proprietatum quibus solida hedris planis inclusa sunt praedita. *Novi Comm. Acad. Sci. Imp. Petropol.* **4** (1752–3, published 1758), 140–160 = *Opera Omnia* (1), Vol. 26, 94–108.

5. KEPLER, J. *Harmonices mundi*, libri V. Lincii Austriae, 1619.

6. WENNINGER, M. *Polyhedron models.* Cambridge University Press, London, 1971.

7. LEGENDRE, A. M. *Éléments de géométrie* (1st edn.). Firmin Didot, Paris, 1794.

8. POINSOT, L. Sur les polygones et les polyèdres. *J. École Polytech.* **4** (Cah. 10) (1810), 16–48.

9. L'HUILIER, S. Démonstration immédiate d'un théorème fondamental d'Euler sur les polyhèdres, et exceptions dont ce théorème est susceptible. *Mém. Acad. Imp. Sci. St. Pétersb.* **4** (1811), 271–301.

10. LISTING, J. B. Der Census räumlicher Complexe oder Verallgemeinerung des Euler'schen Satzes von den Polyëdern. *Abh. K. Ges. Wiss. Göttingen Math. Cl.* **10** (1861–2), 97–182 = Separately as book, Dieterich'sche Verlagshandlung, Göttingen, 1862.

6

The Four-Colour Problem— Early History

A. B. KEMPE (1849–1922)

THE four-colour problem, still unsolved after more than a hundred years, has played a role of the utmost importance in the development of graph theory as we know it today. The origins and early history of the problem will be surveyed in this chapter; in particular, we shall explain why it was thought for some time that the problem had been solved. In Chapter 9 we shall continue the story with an account of the progress made in the first part of the twentieth century.

The origin of the four-colour problem

The first written reference to the four-colour problem occurs in a letter, dated 23 October 1852, sent to Sir William Rowan Hamilton by Augustus DE MORGAN, Professor of Mathematics at University College, London. In this letter, preserved in the manuscript collection of Trinity College, Dublin, he related that he had been asked about the problem by one of his students:

> A student of mine asked me today to give him a reason for a fact which I did not know was a fact—and do not yet. He says that if a figure be anyhow divided and the compartments differently coloured so that figures with any

portion of common boundary *line* are differently coloured—four colours may be wanted, but not more—the following is the case in which four *are*

ABC & D are names of colours

wanted. Query cannot a necessity for five or more be invented. As far as I see at this moment if four *ultimate* compartments have each boundary line in common with one of the others, three of them inclose the fourth and prevent any fifth from connexion with it. If this be true, four colours will colour any possible map without any necessity for colour meeting colour except at a point.

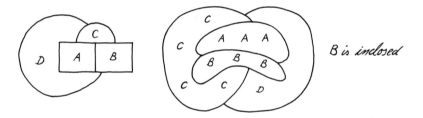

B is inclosed

Now it does seem that drawing three compartments with common boundary ABC two and two you cannot make a fourth take boundary from all, except by inclosing one. But it is tricky work, and I am not sure of the convolutions—what do you say? And has it, if true, been noticed? My pupil says he guessed it in colouring a map of England. The more I think of it, the more evident it seems. If you retort with some very simple case which makes me out a stupid animal, I think I must do as the Sphynx did...

The student who had questioned De Morgan was Frederick Guthrie. In 1880, when the problem had become well known, he published a note [1] which revealed that the originator of the question was his brother, Francis GUTHRIE:

Some thirty years ago, when I was attending Professor De Morgan's class, my brother, Francis Guthrie, who had recently ceased to attend them (and who is now professor of mathematics at the South African University, Cape Town), showed me the fact that the greatest necessary number of colours to be used in colouring a map so as to avoid identity of colour in lineally contiguous districts is four. I should not be justified, after this lapse of time, in trying to give his proof, but the critical diagram was as in the margin.

With my brother's permission I submitted the theorem to Professor De Morgan, who expressed himself very pleased with it; accepted it as new; and, as I am informed by those who subsequently attended his classes, was in the habit of acknowledging whence he had got his information.

If I remember rightly, the proof which my brother gave did not seem altogether satisfactory to himself; but I must refer to him those interested in the subject.

Frederick Guthrie's account of his brother's work is not quite clear, but it is fairly certain that Francis had noticed that four colours are sometimes necessary (as shown by the marginal diagram), and conjectured that this number is always sufficient. Consequently, we are justified in saying that Francis Guthrie was the first person to state the 'four-colour conjecture'. But he published nothing on the problem, and Hamilton, to whom De Morgan had appealed, was obviously not interested. The reply which Hamilton sent to De Morgan on 26 October 1852 stated simply: 'I am not likely to attempt your "quaternion of colours" very soon'.

Before we continue with the story, we should dispose of a popular myth. At one time it was often claimed that the four-colour problem had been invented by A. F. Möbius in 1840; this claim is false, although Möbius did discuss a problem which is superficially similar. In the next chapter we shall describe Möbius' problem, and explain how the confusion arose.

The credit for spreading the four-colour problem belongs to De Morgan; his curiosity had been aroused by Guthrie's question, and he spoke of it to other mathematicians and students, so that the problem became part of mathematical folk-lore. One person who heard about it was the American logician and philosopher, C. S. Peirce. In an unpublished draft (now in the Houghton Library, Harvard University) he claims to have presented an attempt to prove that four colours are sufficient, during the 1860s, to a mathematical society at Harvard. Another manuscript, reliably dated October 1869, mentions the map-colouring problem in connection with his 'logic of relatives'. Peirce retained a lifelong interest in the problem, and some indication of his work may be found in the volumes of his mathematical manuscripts [2], edited by Carolyn Eisele.

Cayley was another who became interested in the problem†. At a meeting of the London Mathematical Society on 13 June 1878 [3], [4], he asked whether it had been proved that four colours are sufficient; soon

---

† Both Cayley [6A] and Kempe [6B] later remarked that the four-colour conjecture had been mentioned 'somewhere' by De Morgan, and it is possible that they intended to cite a published reference, rather than a verbal statement. This interpretation is corroborated in an unpublished manuscript of Peirce, which states that De Morgan had mentioned the conjecture in the journal *Athenaeum*. The authors of this book have searched earnestly in that journal, and in others to which De Morgan contributed regular articles, but nothing of relevance has been found. If this reference does exist, it would be significant, as the first publication of the four-colour conjecture.

afterwards, he sent a short paper [6A] on the question to the Royal Geographical Society. In that paper he attempted to explain in simple terms where the difficulty lies. From his account we infer that there is no good reason to assume, at the outset, that any finite number of colours will always be sufficient—it might be possible to construct maps requiring an arbitrarily large numbers of colours.

6A

A. CAYLEY
ON THE COLOURING OF MAPS

Proceedings of the Royal Geographical Society **1** (1879), 259–261.

The theorem that four colours are sufficient for any map, is mentioned somewhere by the late Professor De Morgan, who refers to it as a theorem known to map-makers. To state the theorem in a precise form, let the term "area" be understood to mean a simply or multiply connected area: and let two areas, if they touch along a line, be said to be "attached" to each other; but if they touch only at a point or points, let them be said to be "appointed" to each other. For instance, if a circular area be divided by radii into sectors, then each sector is attached to the two contiguous sectors, but it is appointed to the several other sectors. The theorem then is, that if an area be partitioned in any manner into areas, these can be, with four colours only, coloured in such wise that in every case two attached areas have distinct colours; appointed areas may have the same colour. Detached areas may in a map represent parts of the same country, but this relation is not in anywise attended to: the colours of such detached areas will be the same, or different, as the theorem may require.

It is easy to see that four colours are wanted; for instance, we have a circle divided into three sectors, the whole circle forming an *enclave* in another area; then we require three colours for the three sectors, and a fourth colour for the surrounding area: if the circle were divided into four sectors, then for these two colours would be sufficient, and taking a third colour for the surrounding area, three colours only would be wanted; and so in general according as the number of sectors is even or odd, three colours or four colours are wanted. And in any tolerably simple case it can be seen that four colours are sufficient. But I have not succeeded in obtaining a general proof: and it is worth while to explain wherein the difficulty consists. Supposing a system of n areas coloured according to the theorem with four colours only, if we add an $(n+1)$th area, it by no means follows that we can *without altering the original colouring* colour this with one of the four colours. For instance, if the original colouring be such that the four colours all present themselves in the exterior boundary of the n areas, and if the new area be an area enclosing the n areas, then there is not any one of the four colours available for the new area.

The theorem, if it is true at all, is true under more stringent conditions. For instance, if in any case the figure includes four or more areas meeting in a point (such as the sectors of a circle), then if (introducing a new area) we place at the point a small circular area, cut out from and attaching itself to each of the original sectorial areas, it must according to the theorem be possible with four colours only to colour the new figure; and this implies that it must be possible to colour the original figure so that only three colours (or it may be two) are used for the sectorial areas. And in precisely the same way (the theorem is in fact really the same) it must be possible to colour the original figure in such wise that only three colours (or it may be two) present themselves in the exterior boundary of the figure.

But now suppose that the theorem *under these more stringent conditions* is true for n areas: say that it is possible with four colours only, to colour the n areas in such wise that not more than three colours present themselves in the external boundary: then it might be easy to prove that the $n+1$ areas could be coloured with four colours only: but this would be insufficient for the purpose of a general proof; it would be necessary to show further that the $n+1$ areas could be with the four colours only coloured *in accordance with the foregoing boundary condition*; for without this we cannot from the case of the $n+1$ areas pass to the next case of $n+2$ areas. And so in general, whatever more stringent conditions we import into the theorem as regards the n areas, it is necessary to show not only that the $n+1$ areas can be coloured with four colours only, but that they can be coloured in accordance with the more stringent conditions. As already mentioned, I have failed to obtain a proof.

The 'proof'

We come now to what is probably the most famous fallacious proof in the whole of mathematics. The author of this 'proof' was a London barrister, Alfred Bray KEMPE, who, having studied mathematics under Cayley at Cambridge, had become well known for his work on linkages. On 17 July 1879, it was announced in *Nature* [5] that Kempe had proved the four-colour conjecture. His proof was contained in a paper [6B] which, possibly at Cayley's suggestion, he submitted to the *American Journal of Mathematics*; as we mentioned in Chapter 4, this journal had recently been founded, and Sylvester was its editor-in-chief. An associate editor W. E. Story studied the paper, made some simplifications which were published as an addendum [6], and communicated the material to a meeting of the Scientific Association at Johns Hopkins University on 5 November 1879 [7]. In addition to Sylvester and Story, the staff of Johns Hopkins included C. S. Peirce. Peirce was present at the November meeting, and at the next meeting, in December, he spoke of his own work on the conjecture [8]. The papers of Kempe and Story were published shortly afterwards in the second volume of the *American Journal*.

Although Kempe's argument contained a fallacy, which we shall pin-point in due course, it is nevertheless worthy of careful study. One technique in particular deserves to be mentioned, for it is used in many later studies of the four-colour conjecture and more general colouring problems.

Suppose that we have a map in which all but one of the regions have been coloured. We may assume that the regions surrounding the uncoloured region have been assigned all four available colours, since otherwise we can immediately colour the last region with a colour not used for the surrounding regions; so the portion of the map depicted in Fig. 6.1 might be typical. If we now consider only those regions coloured red or green, then two possible situations can occur: either the regions A and C are joined by a continuous chain of such regions, or there is no such chain. In the latter case, we may interchange the colours red and green in

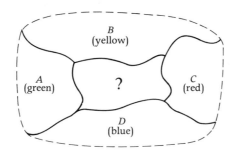

Fig. 6.1.

those red–green regions linked to A, without affecting the colour of C. The result of this interchange is that the regions A and C are now both coloured red, so that the uncoloured region may be coloured green, thereby completing the colouring of the map. However, if there is a chain of red–green regions joining A and C, then the interchange achieves nothing, since all four colours still occur on regions surrounding the uncoloured one. At this stage we may use the fact that our map is drawn on a plane; this means that there cannot be both a red–green chain from A to C and a blue–yellow chain from B to D. So if the interchange argument cannot be applied to A and C, then it can be applied to B and D, thereby reducing the number of surrounding colours to three and enabling the last region to be coloured.

This kind of argument is now referred to as 'the method of Kempe-chains'. It is used several times in [6B], and the mistake occurs in the application of the method to the case when five regions surround the uncoloured one. The reader may care to find the error for himself; alternatively, he may prefer to await the refutation in the final extract [6D] of this chapter.

6B

A. B. KEMPE

ON THE GEOGRAPHICAL PROBLEM OF THE FOUR COLOURS

American Journal of Mathematics **2** (1879), 193–200.

If we examine any ordinary map, we shall find in general a number of lines dividing it into districts, and a number of others denoting rivers, roads, etc. It frequently happens that the multiplicity of the latter lines renders it extremely difficult to distinguish the boundary lines from them. In cases where it is important that the distinction should be clearly marked, the artifice has been adopted by map-makers of painting the districts in different colours, so that the boundaries are clearly defined as the places where one colour ends and another begins; thus rendering it possible to omit the boundary lines altogether. If this clearness of definition be the sole object in view, it is obviously unnecessary that non-adjacent districts should be painted different colours; and further, none of the clearness will be lost, and the boundary lines can equally well be omitted, if districts which merely meet at one or two points be painted the same colour. (See Fig. [6.2].)

This method of definition may of course be applied to the case of any surface which is divided into districts. I shall, however, confine my investigations primarily to the case of what are known as simply or singly connected surfaces, *i.e.* surfaces such as a plane or sphere, which are divided into two parts by a circuit, only referring incidentally to other cases.

If, then, we take a simply connected surface divided in any manner into districts, and proceed to colour these districts so that no two adjacent districts shall be of the same colour, and if we go to work at random, first colouring as many districts as we can with one colour and then proceeding to another colour, we shall find that we require a good many different colours; but, by the use of a little care, the number may be reduced. Now, it has been stated somewhere by Professor De Morgan that it has long been known to map-makers as a matter of experience—an experience however probably confined to comparatively simple cases—that *four* colours will suffice in any case. That four colours may be necessary will be at once obvious on consideration of the case of one district surrounded by three others, (see Fig. [6.3]), but that four colours will suffice in all cases is a fact which is by no means obvious, and has rested hitherto, as far as I know, on the experience I have mentioned, and on the statement of Professor De Morgan, that the fact was no doubt true. Whether that statement was one merely of belief, or whether Professor De Morgan, or any one else, ever gave a proof of

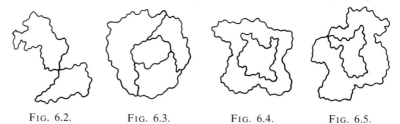

Fig. 6.2.   Fig. 6.3.   Fig. 6.4.   Fig. 6.5.

it, or a way of colouring any given map, is, I believe, unknown; at all events, no answer has been given to the query to that effect put by Professor Cayley to the London Mathematical Society on June 13th, 1878, and subsequently, in a short communication to the Proceedings of the Royal Geographical Society, Vol. I, p. 259. Professor Cayley, while indicating wherein the difficulty of the question consisted, states that he had not then obtained a solution. Some inkling of the nature of the difficulty of the question, unless its weak point be discovered and attacked, may be derived from the fact that a very small alteration in one part of a map may render it necessary to recolour it throughout. After a somewhat arduous search, I have succeeded, suddenly, as might be expected, in hitting upon the weak point, which proved an easy one to attack. The result is, that the experience of the map-makers has not deceived them, the maps they had to deal with, viz: those drawn on simply connected surfaces, can, in every case, be painted with four colours. How this can be done I will endeavour—at the request of the Editor-in-Chief—to explain.

Suppose that we have the surface divided into districts in any way which admits of the districts being coloured with four colours, viz: blue, yellow, red, and green; and suppose that the districts are so coloured. Now if we direct our attention to those districts which are coloured red and green, we shall find that they form one or more detached regions, *i.e.* regions which have no boundary in common, though possibly they may meet at a point or points. These regions will be surrounded by and surround other regions composed of blue and yellow districts, the two sets of regions making up the whole surface. It will readily be seen that we can interchange the colours of the districts in one or more of the red and green regions without doing so in any others, and the map will still be properly coloured. The same remarks apply to the regions composed of districts of any other pair of colours. Now if a region composed of districts of any pair of colours, say red and green as before, be of either of the forms shown in Figures [6.4] and [6.5], it will separate the surface into two parts, so that we may be quite certain that no yellow or blue districts in one part can belong to the same yellow and blue region as any yellow or blue district in the other part. Thus any specified blue district, for example, in one part can, by an interchange of the colours in the yellow and blue region to which it belongs, be converted into a yellow district, whilst any specified yellow district in the other part remains yellow.

Now let us consider the state of things at a point where three or more boundaries and districts meet. It will be convenient to term such a point a *point of concourse*. If three districts meet at the point, they must be coloured with three different colours. If four, they may be coloured with two or three colours only in some cases, but on the other hand they may be coloured with four, as in Fig. [6.6]. If the districts a and c in this case belong to different red and green regions, we can interchange the colours of the districts in one of these regions, and the result will be that the districts a and c will be of the same colour, both red or both green. If a and c belong to the same red and green region, that region will form a ring as in Fig. [6.5], and b will be in one of the parts into which it divides the surface and d in the other, so that the yellow and blue region to which b belongs, will be different from that to which d belongs; if therefore, we interchange the colours in either of these regions, b and d will be of the same colour, both yellow or both blue. Thus we can always reduce the number of colours which meet at the point of concourse of four boundaries to three.

The same thing may be shown in the case of points of concourse where five boundaries meet. The districts meeting at the point may happen to be coloured

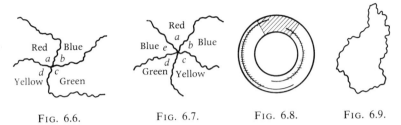

FIG. 6.6.   FIG. 6.7.   FIG. 6.8.   FIG. 6.9.

with only three colours, but they may happen to be coloured with four. Fig. [6.7] shows the only form which the colouring can take in that case, one colour of course occurring twice. If a and c belong to different yellow and red regions, interchanging the colours in either, a and c become both yellow or both red. If a and c belong to the same yellow and red region, see if a and d belong to different green and red regions; if they do, interchanging the colours in either region, a and d become both green or both red. If a and c belong to the same yellow and red region, and a and d belong to the same green and red region, the two regions cut off b from e, so that the blue and green region to which b belongs is different from that to which d and e belong, and the blue and yellow region to which e belongs is different from that to which b and c belong. Thus, interchanging the colours in the blue and green region to which b belongs, and in the blue and yellow region to which e belongs, b becomes green and e yellow, a, c and d remaining unchanged. In each of the three cases the number of colours at the point of concourse is reduced to three.

It will be unnecessary for my purpose to take the case of a larger number of boundaries. Later on, we shall see that we can arrange the colours so that not only will three colours only meet at any given point of concourse, however many boundaries meet there, but also at no point of concourse in the map will four colours appear. It is, however, at present, enough, (and I have proved no more), that if less than six boundaries meet at a point we can always rearrange the colours of the districts so that the number of colours at that point shall only be three.

Before leaving this part of the investigation, I may point out that it does not apply to the case of other surfaces. A glance at Fig. [6.8], which represents an anchor ring, will show that a ring-shaped district, $a\,a$, if it *clasps* the surface, does not divide it into two parts, so that the foregoing proof fails. In fact, six colours may be required to colour an anchor ring. For, if two *clasping* boundaries be described so as to divide the ring into two bent cylindrical portions, and if each portion be divided into three parts by longitudinal boundaries, a, b, c being the three parts on one and d, e, f being those on the other, so that a abuts on d and e at one end, and on e and f at the other; b abuts on e and f at one end, and on f and d at the other; c abuts on f and d at one end, and on d and e at the other, then a, b, c, d, e, f must all be of different colours.

Returning to the case of the simply connected surface, and putting aside for the moment the question of colouring, let us consider some points as to the structure of the map on its surface. This map can have in it *island-districts* having one boundary (Fig. [6.9]); and *island-regions* (Fig. [6.10]), composed of a number of districts; also, *peninsula-districts*, having one boundary and one point of concourse (Fig. [6.11]); and *peninsula-regions* (Fig. [6.12]); *complex-districts*, which have islands and peninsulas in them; and *simple-districts* which have none, and

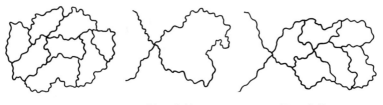

FIG. 6.10.   FIG. 6.11.   FIG. 6.12.

have as many boundaries as points of concourse (Fig. [6.13]). It should also be noticed that, with the exception of those boundaries which are endless, such as that in Figure [6.9], and those which have one point of concourse such as that in Figure [6.11], every boundary ends in two points of concourse; and further, that every boundary belongs to two districts.

Now, take a piece of paper and cut it out to the same shape as any simple-island- or peninsula-district, but rather larger, so as just to overlap the boundaries when laid on the district. Fasten this *patch* (as I shall term it) to the surface and produce all the boundaries which meet the patch, (if there be any, which will always happen except in the case of an island), to meet at a point, (a point of concourse) within the patch. If only two boundaries meet the patch, which will happen if the district be a peninsula, join them across the patch, no point of concourse being necessary. The map will then have one district less, and the numbers of boundaries will also be reduced. Fig. [6.14] shows the district before the patch is put on, the place where it is going to be, being indicated by the dotted line, and Fig. [6.15] shows what is seen after the patch (again denoted by the dotted line) has been put on, and the boundaries have been produced to meet in a point on it. This patching process can be repeated as long as there is a simple district left to operate upon, the patches being in some cases stuck partially over others. If we confine our operations to an island or peninsula, we shall at length get rid of the island or peninsula, and doing this in the case of all the islands and peninsulas, complex-districts will be reduced to simple ones, and can be got rid of by the same process. We can thus, by continually patching, at length get rid of every district on the surface, which will be reduced to a single district devoid of boundaries or points of concourse. The whole map is patched out.

Now, reverse the process, and strip off the patches in the reverse order, taking off first that which was put on last, as each patch is stripped off it discloses a new district, and the map is *developed* by degrees.

Suppose that at any stage of this development, when we have stripped off a

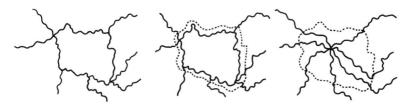

FIG. 6.13.   FIG. 6.14.   FIG. 6.15.

number of patches, there are on the surface

D districts

B boundaries

P points of concourse,

and suppose that after the next patch is stripped off there are

D' districts

B' boundaries

P' points of concourse.

If the patch has no point of concourse on it or line, *i.e.*, if when it is stripped off an island is disclosed,

$$P' = P$$
$$D' = D + 1$$
$$B' = B + 1.$$

If the patch has no point of concourse but only a single line, so that when it is stripped off a peninsula is disclosed,

$$P' = P + 1$$
$$D' = D + 1$$
$$B' = B + 2.$$

If the patch has a point of concourse on it where σ boundaries meet

$$P' = P + \sigma - 1$$
$$D' = D + 1$$
$$B' = B + \sigma.$$

In each case therefore

$$P' + D' - B' - 1 = P + D - B - 1,$$

i.e., at every stage of the development

$$P + D - B - 1$$

has the same value. But at the first stage

$$P = 0$$
$$D = 1$$
$$B = 0.$$

Therefore we always have

$$P + D - B - 1 = 0. \tag{1}$$

That is *in every map drawn on a simply connected surface the number of points of concourse and number of districts are together one greater than the number of boundaries.*

Let d_1, d_2, d_3, etc., denote the number of districts at any stage, which have one, two, three, etc., boundaries, so that

$$D = d_1 + d_2 + d_3 + \ldots,$$

and let p_3, p_4, etc., denote the number of points of concourse, at the same stage of the development, at which three, four, etc., boundaries meet, so that

$$P = p_3 + p_4 + \dots .$$

Then, since every boundary belongs to two districts,

$$2B = d_1 + 2d_2 + 3d_3 + \dots ,$$

and since every boundary ends in two points of concourse, except in the case of continuous boundaries which have no points of concourse, of which let there be β_0, and boundaries round peninsula districts which have one point of concourse, of which let there be β_1, therefore,

$$2B = 2\beta_0 + \beta_1 + 3p_3 + 4p_4 + \dots .$$

Thus, since (1) may be written

$$(6D - 2B) + (6P - 4B) - 6 = 0,$$

we have

$$5d_1 + 4d_2 + 3d_3 + 2d_4 + d_5 - \text{etc.} = 0,$$

the first five terms being the only positive ones. At least one, therefore, of the quantities d_1, d_2, d_3, d_4, d_5, must not vanish, *i.e. every map drawn on a simply connected surface must have a district with less than six boundaries.*

It may readily be seen that this proof applies equally well to an *island-region* or *peninsula-region* as to the whole map. The result is, that we can patch out any simply-connected map, never putting a patch on a district with more than five boundaries. Consequently, if we develop a map so patched out, since each patch, when taken off, discloses a district with less than six boundaries, not more than five boundaries meet at the point of concourse on the patch. Of course districts which, when first disclosed, have only five boundaries may ultimately have thousands.

Returning to the question of colour, if the map at any stage of its development, can be coloured with four colours, we can arrange the colours so that, at the point of concourse on the patch next to be taken off, where less than six boundaries meet, only three colours shall appear, and, therefore, when the patch is stripped off, only three colours surround the disclosed district, which can, therefore, be coloured with the fourth colour, *i.e.* the map can be coloured at the next stage. But, at the first stage, one colour suffices, therefore, four suffice at all stages, and therefore, at the last. This proves the theorem and shows how the map may be coloured.

I stated early in the article that I should show that the colours could be so arranged that only three should appear at every point of concourse. This may readily be shown thus: Stick a small circular patch, with a boundary drawn round its edge, on every point of concourse, forming new districts. Colour this map. Only three colours can surround any district, and therefore the circular patches. Take off the patches and colour the uncovered parts the same colour as the rest of their districts. Only three colours surrounded the patches, and therefore only three will meet at the points of concourse they covered.

A practical way of colouring any map is this, which requires no patches. Number the districts in succession, always numbering a district which has less than six boundaries, not including those boundaries which have a district already numbered on the other side of them. When the whole map is numbered, beginning from the highest number, letter the districts in succession with four letters, *a*, *b*, *c*, *d*, rearranging the letters whenever a district has four round it, so that it may have only three, leaving one to letter the district with. When the whole map is lettered, colour the districts, using different colours for districts lettered differently.

Two special cases should be noticed.

(1). If, excluding island and peninsula districts from the computation, every district is in contact with an even number of others along every circuit formed by its boundaries, three colours will suffice to colour the map.

(2). If an even number of boundaries meet at every point of concourse, two colours will suffice. This species of map is that which is made by drawing any number of continuous lines crossing each other and themselves any number of times.

If we lay a sheet of tracing paper over a map and mark a point on it over each district and connect the points corresponding to districts which have a common boundary, we have on the tracing paper a diagram of a "linkage," and we have as the exact analogue of the question we have been considering, that of lettering the points in the linkage with as few letters as possible, so that no two directly connected points shall be lettered with the same letter. Following this up, we may ask what are the linkages which can be similarly lettered with not less than n letters?

The classification of linkages according to the value of n is one of considerable importance. I shall not, however, enter here upon this question, as it is one which I propose to consider as a part of an investigation upon which I am engaged as to the general theory of linkages. It is for this reason also that I have preferred to treat the question discussed in this paper in the manner I have done, instead of dealing with the analogous linkage.

I will conclude with a theorem which can readily be obtained as a corollary to the preceding results. It is one of which I long endeavoured to obtain an independent proof, as a means of solving the four-colour problem. The polyhedra mentioned are to be understood to be simply connected ones. The theorem is this:

"Polyhedra can be added to the faces of any polyhedron so that in the resulting polyhedron (1) the faces are all triangles, (2) the number of edges meeting at every angular point is a multiple of three."

The supposed solution of the four-colour problem was greeted with great enthusiasm. Peirce remarked upon it in a report which appeared in the *Nation* on 25 December 1879 [9]. Cayley and others proposed that Kempe should be made a Fellow of the Royal Society, on the strength of his work on linkages, and he was duly elected. Improved versions of the 'proof', still containing the essential flaw, were quickly published [10], [11]. We must presume that Kempe's argument was generally accepted as correct.

Kempe's improved 'proof' aroused the interest of P. G. Tait, whom we had cause to mention at the end of Chapter 2. On 15 March 1880 he read a paper on the map-colouring problem to the Royal Society of Edinburgh and an abstract [12] of this paper was published in the *Proceedings* of the Society. Tait began by stating that Cayley had told him about the problem some years previously, and that he himself had shown that, in cases where an even number of boundary lines meet at each point, two colours will always suffice. He then sketched some ideas which, he claimed, would lead to alternative proofs of the four-colour conjecture. With hindsight, we are not disposed to treat his claim too seriously.

It was the published abstract of Tait's paper which prompted Frederick Guthrie to publish his account of the origin of the problem in the same *Proceedings* [1]. Immediately following this account is the abstract of another paper by Tait, delivered on 19 July 1880 [6C]. This second paper amplified an idea which he had mentioned in the first one; it deserves careful study, since it contains a mixture of a brilliant idea with a certain amount of nonsense.

Both Cayley and Kempe, in the extracts given above, had pointed out that it is sufficient to prove the four-colour conjecture for the case when three boundary lines meet at each point—in other words, when each vertex of the underlying graph has valency three. We shall use the modern term, **trivalent**, to describe such a graph or map. Tait noticed that a four-colouring of the regions of a trivalent map is equivalent to an assignment of three colours to the edges of the map, in such a way that all three colours appear at each vertex. The procedure which he gave in [6C] is illustrated in Fig. 6.16; this procedure, and the resulting equivalence with the four-colour conjecture, are essentially correct, but the other remarks in [6C] must be viewed with scepticism.

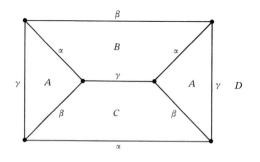

FIG. 6.16.

6C

P. G. TAIT
[REMARKS ON THE COLOURING OF MAPS]

Proceedings of the Royal Society of Edinburgh **10** (1878–80), 729.

In a paper read to the Society on 15th March last, I gave a series of proofs of the theorem that four colours suffice for a map. All of these were long, and I felt that, while more than sufficient to prove the truth of the theorem, they gave little insight into its real nature and bearings. A somewhat similar remark may, I think, be made about Mr Kempe's proof.

But a remark incidentally made in the abstract of my former paper has led me to a totally different mode of attacking the question, which puts its nature in a clearer light. I have therefore withdrawn my former paper, as in great part superseded by the present one.

The remark referred to is to the effect that, if an even number of points be joined, so that three (and only three) lines meet in each, these lines may be coloured with *three* colours only, so that no two conterminous lines shall have the same colour. (When an odd number of the points forms a group, connected by *one* line only with the rest, the theorem is not true.)

This follows immediately from the main theorem when it is applied to a map in which the boundaries meet in threes (and the excepted case cannot then present itself). For we have only to colour such a map with the colours A, B, C, D. Then if the common boundaries of A and B, as also of C and D, be coloured α; those of A and C, and of B and D, β; and those of A and D, and of B and C, γ, it is clear that the three boundaries which meet in any one point will have the three colours α, β, γ.

The proof of the elementary theorem is given easily by induction; and then the proof that four colours suffice for a map follows almost immediately from the theorem, by an inversion of the demonstration just given.

We escape the excepted case by taking the points as the summits of a polyhedron, all of which are trihedral; and when the figure is a pentagonal dodecahedron the theorem leads to Hamilton's *Icosian Game*.

Tait believed that Kempe had proved the four-colour conjecture, and so he had some excuse for thinking that his 'elementary theorem' was true in consequence. But Kempe was wrong, and Tait's claim that a direct proof of the elementary theorem is easily given by induction was also wrong—indeed, the new problem is just as difficult as the original one. The full version of Tait's second paper was eventually published in the *Transactions of the Royal Society of Edinburgh*; it contains the same mixture of sense and nonsense, including the statement that Hamilton took the idea for the Icosian Game from Kirkman [13, p. 657]. We shall return to these muddied waters, suitably prepared, in Chapter 10.

Heawood and the five-colour theorem

Throughout the 1880s the four-colour 'theorem' was regarded as an established fact, and its fame spread far and wide. For example, Lewis Carroll (C. L. Dodgson) turned it into a game, in which one player drew a map for a second player to colour [14].

There were also contributions from rather more unexpected sources. In 1886, the Headmaster of Clifton College, an English boys' school, posed the problem as a 'challenge' to the school:

> In colouring a plane map of counties, it is of course desired that no two counties which have a common boundary should be coloured alike; and it is 'found, on trial, that four colours are always sufficient, *whatever the shape or number* of the counties or areas may be. Required, a good proof of this. Why *four*? Would it be true if the areas are drawn so as to cover a whole sphere?
>
> Solutions to be sent to the Head Master on or before Dec. 1, with mottos, and envelopes bearing the motto outside and the name inside.
>
> No solution may exceed one page, 30 lines of MS., and one page of diagrams.

The Headmaster's challenge was published in the *Journal of Education* [15], with an editorial comment which suggested that: 'Perhaps it will tempt some of our correspondents to send us a solution'. This hope was to be fulfilled. The Bishop of London (Frederick Temple, later Archbishop of Canterbury) allowed his mind to wander while attending a meeting; oblivious to the speaker, he wrote out his own 'proof', which later appeared in the same *Journal* [16]. In 1906 J. C. Wilson [17] pointed out Temple's error.

The fallacy in Kempe's proof was first revealed in 1890. In the introduction to his paper on the 'map-colour theorem', Percy John HEAWOOD almost apologized for having found a mistake in Kempe's argument. Our extract [6D] contains those parts of his paper in which Heawood demonstrated the fallacy and showed, by a modified Kempe-chain argument, that five colours are always sufficient for colouring a plane map. In the same paper he investigated the colouring of maps on other surfaces, a topic which we shall discuss at length in the next chapter.

6D

P. J. HEAWOOD
MAP-COLOUR THEOREM

Quarterly Journal of Pure and Applied Mathematics **24** (1890), 332–338.

The Descriptive-Geometry Theorem that any map whatever can have its divisions properly distinguished by the use of but four colours, from its generality and intangibility, seems to have aroused a good deal of interest some few years ago when the rigorous proof of it appeared to be difficult if not impossible, though no case of failure could be found. The present article does not profess to give a proof of this original Theorem; in fact its aims are so far rather destructive than constructive, for it will be shown that there is a defect in the now apparently recognised proof.

*   *   *   *   *

Of five divisions surrounding another it will be easily seen that there must be some pair touching each other nowhere; when we reduce the map by removing the division which the five surround, we may, instead of merely making the 5 divisions close up and touch (as in Fig. [6.17]), throw such a pair of them into one with the division removed, as *e.g.*, 2 and 5 in Fig. [6.18]. The new map thus formed, coloured with five colours, will show how the original map also could be coloured with five colours; two of the five divisions being in the reduced map portions of one division, and therefore necessarily of the same colour, so that four colours only will appear within their boundaries, and the fifth will be available for the division which they touch. Reduction being continued as before, this proves that five colours are sufficient, and a similar method applied to a division touched by four others would reduce the map on the supposition of but four colours. However, the required problem is to show that 4 colours are sufficient for the map's colouring to follow from a reduced map, even where a five-contact division is the lowest. Mr. Kempe's device aims at meeting this case by showing that, if a map can be done in 4 colours apart from such a division, and should all four occur round it, as in Fig. [6.17], we can then so transpose the colours as to remove from those in contact with one colour, with which it can therefore be coloured. Throughout a *red–blue* region, for instance, that is, throughout the whole of a continuous aggregate of red and blue divisions, the red and the blue may be obviously transposed without any alteration in the rest of the map. Now suppose that 2 and 5 (Fig. [6.17]) belong to different *blue–yellow* regions. The transposition of blue and yellow throughout one of the two regions will remove one of these colours from occurring round the division for which a colour is required. Similarly, a colour may be removed, if 2 and 4 belong to different *blue–green* regions. However, 2 and 5 may belong to the same *blue–yellow*, and 2 and 4 to the same *blue–green* region. In this case, however, 1 and 4 are necessarily cut off from belonging to the same *red–green,* and 3 and 5 from belonging to the same *red–yellow* region, as seen in Fig. [6.19]. Therefore—Mr. Kempe says—the transposition of colours throughout 1's *red–green* and 3's *red–yellow* regions will each remove a red, and what is required is done. If this were so, it would at once

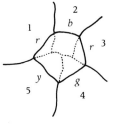

FIG. 6.17.   FIG. 6.18.

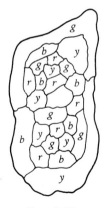

FIG. 6.19.

lead to a proof of the proposition in question; by showing that the colouring of a map in 4 colours could *always* be made to depend on the colouring of a reduced map in 4 colours, it would prove that 4 colours were sufficient for any map. But, unfortunately, it is conceivable that though *either* transposition would remove a red, *both* may not remove both reds. Fig. [6.19] is an actual exemplification of this possibility, where either transposition prevents the other from being of any avail, by bringing the red and the other division into the same region; so that Mr. Kempe's proof does not hold, unless some modifications can be introduced into it to meet this case of failure. There is another simpler proposition not yet noticed, that a map can be coloured with but 3 colours if all its divisions are touched by an even number of others. The proof of this is not difficult, but it appears to shed no light on the main proposition, which, in spite of all attempts to prove it or disprove it, seems to have been hitherto attempted with very partial success.

Heawood's discovery was reported to the London Mathematical Society by Kempe himself [18]. He admitted his error, and conceded that he had not been able to remedy the defect. Later, on 9 April 1891, he discussed the question at a meeting of the Society [19].

It is perhaps natural that the refutation of an accepted proof should be greeted with a certain lack of enthusiasm. Whatever the cause, we find no complimentary references to Heawood in the popular journals, and no record of honours granted to him. (Many years later he was to receive an honorary degree and the Order of the British Empire—for his efforts towards the restoration of Durham Castle.)

The obscurity in which Heawood's paper remained may be inferred from the pages of *L'Intermédiaire des Mathématiciens,* a journal founded in 1894 for the dissemination of problems and solutions. One of its first problems, set by P. Mansion in February 1894 [20], was just a statement of the four-colour conjecture. Almost immediately, replies were published from H. Delannoy and A. S. Ramsey [21], stating that the problem had been solved and giving references to the work of Kempe and Tait;

the fourth volume of Lucas' book [22], which gave an account of Kempe's work, was also cited. There was no mention of the fact that Heawood had refuted the purported solution. Further correspondence ensued, but it soon became bogged down in arguments about the work of Tait (see Chapter 10). Even in 1896, when C. de la Vallée Poussin [23] pointed out the error in Kempe's argument, as given by Lucas, there was still no mention of Heawood.

This point in our book marks the end of the 'Age of Innocence' for graph theory. Until now we have been discussing articles and papers using specific techniques which, with hindsight, we recognize as belonging to graph theory. From now on, we shall be dealing mainly with papers which set out to build up the theory as a mathematical discipline in its own right. In particular, in Chapter 9 we shall discuss the theory of colouring which was developed in the face of the challenge of the four-colour conjecture.

References

1. GUTHRIE, FREDERICK. Note on the colouring of maps. *Proc. Roy. Soc. Edinb.* **10** (1880), 727–728.
2. PEIRCE, C. S. *New elements of mathematics*, 4 vols. Mouton, Den Haag.
3. [Report of meeting, 13 June 1878]. *Proc. London Math. Soc.* **9** (1877–8), 148.
4. [Report of London Mathematical Society meeting]. *Nature* **18** (1878), 294.
5. [Notes]. *Nature* **20** (1879), 275.
6. STORY, W. E. Note on the preceding paper. *Amer. J. Math.* **2** (1879), 201–204.
7. [Report of meeting of Scientific Association, 5 November 1879]. *Johns Hopkins Univ. Circ.* **1**, No. 2 (Jan. 1880), 16.
8. [Report of meeting of Scientific Association, 3 December 1879]. *Johns Hopkins Univ. Circ.* **1**, No. 2 (Jan. 1880), 16.
9. PEIRCE, C. S. [Untitled abstract]. *Nation* No. 756 (1879), 440.
10. KEMPE, A. B. [Untitled abstract]. *Proc. London Math. Soc.* **10** (1878–9), 229–231.
11. KEMPE, A. B. How to colour a map with four colours. *Nature* **21** (1879–80), 399–400.
12. TAIT, P. G. On the colouring of maps. *Proc. Roy. Soc. Edinb.* **10** (1878–80), 501–503.
13. TAIT, P. G. Note on a theorem in the geometry of position. *Trans. Roy. Soc. Edinb.* **29** (1880), 657–660 = *Sci. Papers*, Vol. 1, 408–411.
14. COLLINGWOOD, S. D. *The life and letters of Lewis Carroll (Rev. C. L. Dodgson)*. Nelson, London, 1898.
15. [Note by the Headmaster of Clifton College]. *J. Educ.* **9** (1887), 11–12.
16. [Note by the Headmaster of Clifton College]. *J. Educ.* **11** (1889), 277.
17. WILSON, J. C. On a supposed solution of the 'four-colour problem'. *Math. Gaz.* **3** (1904–6), 338–340.
18. [Appendix]. *Proc. London Math. Soc.* **21** (1889–90), 456.
19. [Report of Meeting, 9 April 1891]. *Proc. London Math. Soc.* **22** (1890–1), 263.
20. MANSION, P. Question 51. *Interméd. Math.* **1** (1894), 20.
21. DELANNOY, H. and RAMSEY, A. S. [Réponse à question 51]. *Interméd. Math.* **1** (1894), 192.
22. LUCAS, É. *Récréations mathématiques*, Vol. 4. Gauthier-Villars, Paris, 1894.
23. DE LA VALLÉE POUSSIN, C. [Problème des quatre couleurs-deuxième réponse]. *Interméd. Math.* **3** (1896), 179–180.

7

Colouring Maps on Surfaces

P. J. Heawood (1861–1955)

In his paper [6B] on the four-colour problem, Kempe remarked that
maps on surfaces more complicated than a sphere may require more than
four colours. Indeed, he gave an example of a map on a torus which
needs six colours. Later writers, from Heawood onwards, turned to the
general problem in the hope that more general methods might illuminate
the problem for the special case of maps on a sphere. It is true that
general techniques were developed, and that these led eventually (in
1967) to a complete solution of the problem on higher surfaces, but
surprisingly the problem of maps on spheres remains unsolved. In this
chapter we shall survey some of the early work on the colouring of maps
on surfaces other than the sphere.

The chromatic number of a surface

In Chapter 5 we described a family of closed surfaces, the 'spheres with
handles', the simplest examples of which are the sphere itself, the torus
(Fig. 5.8), and the double-torus (Fig. 5.10); a sphere with p handles is
said to have genus p. We also remarked that it is possible to construct
other closed surfaces which have only one 'side', unlike the surfaces of
genus p which have two sides. Later in this chapter we shall give a full
explanation of the concept of one-sidedness; for the moment it is sufficient to consider only two-sided surfaces, although our arguments will
often apply to both cases.

For a given map M on a closed surface S, the symbol $\chi(M)$ will denote the **chromatic number** of M—this is defined to be the smallest number of colours which suffices for colouring the faces of M, subject to the usual constraint that faces which have an edge in common shall be coloured differently. At first sight it might seem possible that, for a given surface S, we could construct a map M on S whose chromatic number is as large as we please. However, it turns out that the opposite is true: there is an upper bound for $\chi(M)$ which depends only on the surface S. This fact was first published by Heawood in 1890, in the same paper in which he demolished Kempe's arguments. Heawood's proof is rather complicated, and so, before embarking upon it, we shall give a more straightforward version.

We already know one thing which is invariant for all maps on a given surface. If N_v, N_e, and N_f denote the numbers of vertices, edges, and faces of a map M on a surface S, then the Euler characteristic defined by

$$\eta(M) = N_v - N_e + N_f$$

depends only on S, so that we are justified in writing $\eta(S)$ for its constant value. The first step in the colouring problem is to use this formula to show that every map on S must have at least one face with a 'small' number of neighbours.

There is no effective loss of generality in assuming that all our maps have two simple properties; we may assume, first, that every edge bounds two faces, and secondly, that the valency of each vertex is at least three. Under these assumptions, simple counting arguments show that

$$3N_v \leqslant 2N_e \quad \text{and} \quad \alpha N_f = 2N_e,$$

where α is the average number of edges surrounding a face. Substituting for N_e and N_v in the definition of the Euler characteristic, we obtain the bound

$$\alpha \leqslant 6\left(1 - \frac{\eta(S)}{N_f}\right).$$

In order to convert this bound for α into an absolute one, depending only on the surface S, we consider the quadratic expression

$$Q(t) = t^2 - 7t + 6\eta(S).$$

From the fact that $\eta(S) \leqslant 2$ for all surfaces, it follows that Q has two real zeros, and that the larger one (β, say) is strictly positive. Thus

$$\beta^2 - 7\beta + 6\eta(S) = 0,$$

and so

$$6\left(1 - \frac{\eta(S)}{\beta}\right) = \beta - 1.$$

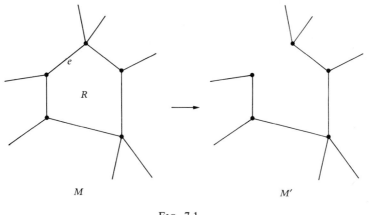

Fɪɢ. 7.1.

Comparing this with the bound for α, we see that, provided $\eta(S) \leqslant 0$ and $N_f \geqslant \beta$, the average number of edges surrounding a face must satisfy

$$\alpha \leqslant \beta - 1.$$

Let h denote the integer part of β. There must be at least one face with not more than the average number of edges, and so we have a face with not more than $h-1$ edges. In the following discussion we emphasize the dependence on S by writing $h(S)$ instead of h. We have proved that every map on S has a face with not more than $h(S)-1$ edges, given that $\eta(S) \leqslant 0$ and that the total number of faces exceeds $h(S)$.

These remarks enable us to prove, by induction on the number of faces, that

$$\chi(M) \leqslant h(S)$$

for all maps M on a surface S with $\eta(S) \leqslant 0$. For, suppose first that $N_f \leqslant h(S)$. Since we need at most N_f colours for N_f faces, it follows that $\chi(M) \leqslant h(S)$, and we may start the induction. If $N_f > h(S)$, then the results of the previous paragraph imply that M has a face R with not more than $h(S)-1$ neighbours. We construct a new map M' on S by removing an edge e which bounds the face R (see Fig. 7.1). The map M' has N_f-1 faces, and we make the induction hypothesis that it is coloured with not more than $h(S)$ colours. Since there are at most $h(S)-1$ faces adjacent to R in M, there is a spare colour not used on the corresponding faces of M'; so we may reinstate the edge e, assign to R the spare colour, and thereby obtain a colouring of M with not more than $h(S)$ colours. Thus the result is proved.

The foregoing argument applies to any map M on a surface S with $\eta(S) \leqslant 0$; using the familiar formula for the larger root of the quadratic equation $Q(t) = 0$, we obtain

$$\chi(M) \leq h(S) = [\tfrac{7}{2} + \tfrac{1}{2}\sqrt{(49 - 24\eta(S))}],$$

where $[y]$ denotes the integer part of y. The number $h(S)$ is often called the **Heawood number** of S. So if $\chi(S)$ is the minimum number of colours which will colour any map on S, we have proved that $\chi(S)$ exists and obtained the explicit upper bound

$$\chi(S) \leq h(S).$$

The number $\chi(S)$ is called the **chromatic number** of S.

It is important to notice what has *not* been proved. Our proof works only if $\eta(S) \leq 0$; this excludes the original case when S is a sphere, and just one other (one-sided) closed surface. In addition, it has been proved only that $h(S)$ colours are sufficient, not that they are necessary. We shall return to this point after Heawood's paper.

The proof given by Heawood [7A] is similar to the one given above, but it does not always proceed in a logical order. For instance, he assumed (without saying so) that all maps are trivalent, derived a formula, and then returned to proving that, in fact, maps can be modified so as to make them trivalent.

7A

P. J. HEAWOOD
MAP-COLOUR THEOREM
Quarterly Journal of Pure and Applied Mathematics **24** (1890), 332–338.

*   *   *   *   *

In dealing with a map on an ordinary singly-connected surface, or, better, on a sphere (for we need only consider complete surfaces, a plane map being equivalent to a portion of a map on a sphere), we have the one definite fact to start with that if there are n divisions, the average number, (A_n), of contacts of a division with the others is $6 - \dfrac{12}{n}$. This is easily proved by induction: if we reduce the number of divisions to $n - 1$ by annihilating a division touched by, say m others, and closing up the divisions round it in any way, we shall now have (fig. [7.2]) $m - 3$ new contacts of these divisions with each other in place of their m contacts with the division which has disappeared. Thus the sum of the numbers of contacts of all the divisions is reduced by 6 (each line of contact belonging to two divisions, and therefore counting twice); the average with n divisions will therefore be $\dfrac{6(n-2)}{n}$, provided the average with $n - 1$ divisions is $\dfrac{6(n-2)-6}{n-1}$ or $\dfrac{6(n-3)}{n-1}$; but the law ob-

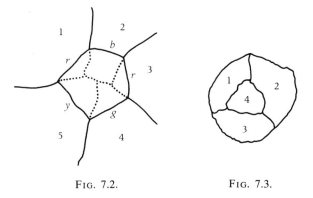

FIG. 7.2.   FIG. 7.3.

viously holds for 4 divisions (fig. [7.3]); therefore it holds for 5 and for 6 and so on universally. It is, indeed, the equivalent of the polyhedron formula $S + F = E + 2$ just in the special case where all the angles are trihedral and $3S = 2E$, so that $\frac{2E}{F} = 6 - \frac{12}{F}$ $\left(\text{as in our formula } A_n = 6 - \frac{12}{n}\right)$, which implies that in a map we only recognise the meeting of three divisions at a point, as is in fact assumed in the above proof.

If four boundaries meet in a point (fig. [7.4]), the supposition may be made that one pair of opposite divisions, 1 and 3, touch; but this must debar the other pair, 2 and 4, from touching, and we have to consider the point as equivalent to two meeting points of three divisions joined by an infinitesimal boundary line, as shown in the figure. It should also be noticed that if a division touches another at two separate places (fig. [7.5]) we must reckon two distinct contacts. With this understanding $A_n = 6 - \frac{12}{n}$ universally, at least where n is not less than 3. The only exception is where a division or divisions are entirely surrounded by one other, where we may treat the enclosed divisions as forming a separate map; or we may observe generally that all such cases of degeneracy give a *lower* not a *higher* number of contacts than there would otherwise be for that number of divisions, which it will be seen does not interfere with any of the deductions presently to be made from the formula.

For more highly connected surfaces a similar law holds and the proof is similar. On an anchor-ring we find $A_n = 6$, generalising by induction from a special simple case (fig. [7.6]) just as before. So generally $A_n = 6\left(1 + \frac{k}{n}\right)$, where k is a positive

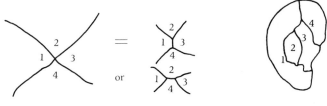

FIG. 7.4.   FIG. 7.5.

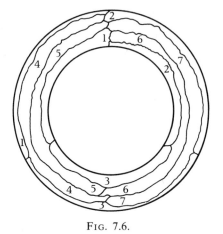

FIG. 7.6.

constant depending on the connectivity of the complete surface on which the map is drawn: degenerate cases leading as before to a lower rather than a higher number of contacts, being indeed equivalent to a repetition of maps on simpler surfaces.

Now the above law gives us at once a superior limit to the number of colours required for a map. For let x be the next integer to the greatest value of A_n, and therefore $= 6$ for an ordinary map; x colours will certainly be sufficient to colour the map with. For it follows immediately, the average number of contacts being less than 6, that there must be some divisions touched by not more than 5 others: we can then annihilate such a division, making those round it close up in the space which it occupied (fig. [7.2]), and obviously the original map can be done in 6 colours if the reduced map can (for whatever the colouring of the divisions around, there would be a colour to spare for the division annihilated); and we can proceed with our reduction till, say, 6 divisions are left, and the truth of the proposition is evident. An inferior limit to the number of colours can also be found. For suppose that we can construct a map of any number of divisions each touching all the rest, it is obvious that at least that number of colours must be necessary. Our formula $A_n = 6 - \dfrac{12}{n}$ shows that the number of such divisions cannot be greater than 4 for an ordinary map; for it cannot be greater than such an integer y as will not make $A_y < y - 1$; for A_y must have at least this value to enable each of y divisions to touch $y - 1$ others; for instance, $A_5 = 3\frac{3}{5}$ shews that we could not have five divisions each touching all the rest which would necessitate $A_5 = 4$ at least. It is obvious that four divisions each touching each can be actually realised (fig. [7.3]), and therefore four colours are necessary; but it is important to notice that verification by reference to an actual figure must not be omitted. It is conceivable that y divisions might each have $y - 1$ contacts, and yet owing to double contacts they might fail each to touch all the rest; or at any rate that this might be the case although $A_y = y - 1$, and then y colours might not be necessary for the y divisions, as, in fact, four colours are not necessary for the four divisions in the special case of (fig. [7.5]). For an ordinary map we have so far only proved that four colours are necessary and that six are sufficient, but for a map described on an anchor-ring we have seen that $A_n = 6$, so that $x = 7$; therefore seven colours are always sufficient for such a map; while $A = y - 1$ when $y = 7$; and therefore seven colours are presumably necessary, as there are just sufficient contacts for seven divisions each to touch each; also (fig. [7.6]) gives the required verification that such a case can be realised; thus for a map on such a surface $x = y = 7$, and we can prove at once that seven colours are both necessary and sufficient. Moreover, for higher kinds of surface in general, A_n being $= 6\left(1 + \dfrac{k}{n}\right)$, which diminishes as n increases, we have a slight difference in the determination of x, which need not now be greater than the greatest value of A_n, but only greater than A_{x+1}. For this ensures $A_n < x$ for all values of n greater than x, which is of course sufficient to ensure that there must be some divisions touched by fewer than x others. (A higher average when n is lower than this can only be accounted for by duplicate and triplicate contacts of the same divisions.) The condition for x being now $A_{x+1} < x$, and for y as before A_y not $< y - 1$, x and y necessarily have a common value, namely the greatest integer [which] will not make

$$A_x \quad \text{or} \quad 6\left(1 + \frac{k}{x}\right) < x - 1,$$

which gives $x^2 - 7x$ not $> 6k$, i.e. x not $> \frac{1}{2}\{7 + \sqrt{(24k + 49)}\}$. This or the next less integer is therefore the number of colours necessary and sufficient for a surface where k is positive or zero, and a single surface such as a sphere is the only one where x and y are not necessarily equal and where we consequently do not get an immediate definite solution of the problem,—apart from the verification figure, which we have indeed only given for the case of an anchor ring, but for more highly connected surfaces it will be observed that there are generally contacts enough and to spare for the above number of divisions each to touch each.

* * * * *

Neighbouring regions

Heawood clearly stated that in order to prove $\chi(S) = h(S)$ we must construct a map on S which requires $h(S)$ colours. Yet he did so only when S is a torus, dismissing the general case with the claim that 'there are generally contacts enough and to spare'. The first steps towards a comprehensive proof of this difficult result were taken by a German mathematician, Lothar HEFFTER, soon after the publication of Heawood's paper. His idea was to construct maps in which each face is adjacent to every other one, and he called such a map a system of **neighbouring regions**.

In fact the problem of neighbouring regions on the plane was the subject of a puzzle discussed by A. F. Möbius, in his lectures around the year 1840. It was this puzzle, which we shall now describe, that led to the mistaken belief that Möbius had invented the four-colour problem.

There was once a king (said Möbius) with five sons. In his will he stated that after his death the sons should divide the kingdom into five regions in such a way that the boundary of each one should have a frontier line (not

just a point) in common with each of the other four. Can the terms of the
will be satisfied?

The problem asks for a system of five neighbouring regions in the
plane, and one can easily prove, using Euler's formula, that it is insoluble
(see Chapter 8). But how did this simple result come to be confused with
the four-colour problem?

In 1885 R. Baltzer published an account [1] of how the problem of five
neighbouring regions had been suggested to Möbius by his friend at the
University of Leipzig, the philologist, Professor Weiske. At the end of his
paper, he mentioned that Felix Klein had pointed out to him the
similarity between this problem and the four-colour problem. Baltzer's
paper was brought into prominence by a 'Note on the history of the
map-coloring problem' [2], written in 1897 by Isabel Maddison of Bryn
Mawr College. Writing of the four-colour conjecture, she claimed that
'Möbius discussed the question, in a slightly different form, in his lectures
in 1840', and cited Baltzer as a reference. After this, the notion that
Möbius had originated the four-colour conjecture became a part of
mathematical folk-lore, and was mentioned in such popular works as
W. W. Rouse Ball's *Mathematical recreations and essays* [3, p. 223] and
E. T. Bell's *The development of mathematics* [4, p. 606].

However, the logical connection between the map-colouring problem
and the problem of Möbius and Weiske is a one-way relationship. If five
neighbouring regions could be constructed, then five colours would be
needed for colouring the map so formed; but the fact that five such
regions cannot be found proves nothing. Also it seems certain that
Möbius himself saw no relation between his problem and the colouring of
maps. It was not until 1959 that the record was put right, in an article by
H. S. M. Coxeter [5].

We now return to the work of Heffter. He had read Baltzer's article,
and he saw that the concept of neighbouring regions is relevant to the
problem left unfinished by Heawood, since in this instance the one-way
relationship operates in the right direction. In order to colour a map of
neighbouring regions, all the faces must have different colours, and so the
chromatic number of the map is equal to the number of its faces.
Consequently, Heffter's approach involved the construction of $h(S)$
neighbouring regions on a surface S, from which it would follow that
$\chi(S) = h(S)$.

In fact, his paper [7B] is mainly concerned with the closed two-sided
surfaces; we shall use the notation S_p to denote such a surface, where p is
the genus. The Euler characteristic is given by Lhuilier's formula $\eta(S_p) =
2 - 2p$, so that if we write $h(p)$ for $h(S_p)$, Heawood's result becomes

$$\chi(S_p) \leqslant h(p) = [\tfrac{7}{2} + \tfrac{1}{2}\sqrt{(1 + 48p)}],$$

provided that $p \geqslant 1$.

Heffter employed a notable degree of dexterity in his attempts to construct $h(p)$ neighbouring regions on the surface S_p; he not only turned the problem upside down, but inside out as well! His first step was to fix the number n of neighbouring regions and vary the genus p, instead of vice versa. That is, he asked for the least number p such that n neighbouring regions may be constructed on S_p. If p_n denotes this least number, then it is quite easy to apply Euler's formula to show that $p_n \geqslant k(n)$, where $k(n)$ denotes the least integer not less than

$$\frac{(n-3)(n-4)}{12} \qquad (n \geqslant 7);$$

this proof is included in extract [7B]. Furthermore, if the equality $p_n = k(n)$ holds for a particular value of n, then we also have the desired equality

$$\chi(S_p) = h(p),$$

for all values of p in the range $k(n) \leqslant p < k(n+1)$. To see this, we observe that if $p_n = k(n)$ then there are n neighbouring regions on $S_{k(n)}$, whence $\chi(S_{k(n)}) \geqslant n$. So, if $p \geqslant k(n)$, then, by Heawood's theorem,

$$n \leqslant \chi(S_{k(n)}) \leqslant \chi(S_p) \leqslant [H],$$

where $H = \frac{7}{2} + \frac{1}{2}\sqrt{(1+48p)}$. If in addition $p < k(n+1)$, then

$$p < \frac{(n-2)(n-3)}{12}, \quad \text{that is,} \quad n^2 - 5n + (6-12p) > 0.$$

For $p \geqslant 1$ and $n \geqslant 7$ this implies that

$$n > \tfrac{5}{2} + \tfrac{1}{2}\sqrt{(1+48p)} = H - 1.$$

In summary, we have the inequalities $H - 1 < \chi(S_p) \leqslant [H]$, which, since $\chi(S_p)$ is an integer, gives the required result,

$$\chi(S_p) = [H] = h(p).$$

Heffter's second inversion of the problem used the notion of 'duality' for maps. If we are given a map M on a surface S, then we may construct a **dual** map M^* in the following way: place a vertex of M^* inside each face of M, and join two such vertices by an edge whenever the corresponding faces of M have a common edge. (It is convenient to make the edges of M^* cross those of M, so that an explicit one-to-one correspondence is set up.) An example of this construction, in which the surface S is chosen to be the plane, is shown in Fig. 7.7. In general, the numbers of vertices, edges, and faces of M^* are given in terms of the numbers for M by the equations

$$N_v^* = N_f, \qquad N_e^* = N_e, \qquad N_f^* = N_v.$$

Also, if M is connected, then M^{**} (the dual of M^*) is just M, so that we have a true duality.

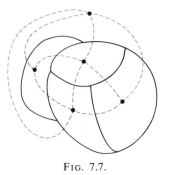

F IG. 7.7.

The idea of duality was mentioned by Kempe at the end of his 1879 paper [6B], and it had been used before that in the study of stresses in frameworks, notably by J. C. Maxwell [6], [7], who is best known for his work in electromagnetic theory. But the concept had originated in the study of polyhedra, as may be seen by referring to the first paragraph of our extract from Lhuilier's work [5C]. Indeed, the duality between the cube and the octahedron, and between the dodecahedron and the icosahedron, goes back to the so-called 'fifteenth book of Euclid' (see [8]); the origin of this book is not known with any certainty, but it is much later than Euclid, and probably dates from the sixth century A.D.

The dual of a map of neighbouring regions is called, in Heffter's terminology, a system of **neighbouring points**. In such a system each vertex (point) is joined to all the others, so that the graph formed by the vertices and edges is a complete graph. Thus the dual form of Heffter's problem asks for the least genus p such that the complete graph K_n can be drawn on S_p, and this is the form in which the problem is treated in [7B]. However, there are occasional, rather confusing references to the problem in the dual 'neighbouring regions' formulation.

7B

L. HEFFTER
ÜBER DAS PROBLEM DER NACHBARGEBIETE
[On the problem of neighbouring regions]

Mathematische Annalen **38** (1891), 477–508.

*   *   *   *   *

We next enquire after the minimal value of the genus of a surface which does admit n neighbouring points, and call this number p_n; that is to say it will be supposed that on a surface of genus p_n there really exist n neighbouring points, but not on one of lower genus. Then it is easy to determine a lower bound for p_n.

By virtue of the n neighbouring points and their $\frac{1}{2}n(n-1)$ connecting lines, a system of n points or vertices, $\frac{1}{2}n(n-1)$ lines or edges, and a number (F) of faces or bounded regions are formed on the closed surface. The last are all simply-connected; for were a multiply-connected piece to arise thereby, then we could place on it a closed contour not contractible to a point, which would not cut up the undivided surface. Thus the genus of the surface could be reduced by one, and there would also be n neighbouring points on it, contrary to hypothesis. For the number F of these simply-connected faces, we obtain the upper bound

$$F \leq \frac{n(n-1)}{3};$$

this is because around each of the n points lie $n-1$ faces, which certainly are at least triangles, since if digons were to be found two of the n points would be doubly joined with each other. Now we apply to the system of vertices, edges and faces of the surface of genus p_n the generalized Euler polyhedral theorem, thus obtaining

$$2p_n - 2 = \frac{n(n-1)}{2} - n - F,$$

$$2p_n - 2 \geqq \frac{n(n-1)}{2} - n - \frac{n(n-1)}{3},$$

$$p_n \geqq \frac{(n-3)(n-4)}{12},$$

and since in every case p_n must be an integer, we can write

(A)
$$p_n \geqq \frac{(n-3)(n-4)+2\alpha_n}{12},$$

where $2\alpha_n$ is the smallest positive integer which makes the numerator divisible by 12. It is very easily seen that this number can only be an *even* one, and furthermore, that α_n can only have one of the values 0, 2, 3, 5.

8. For the lowest values of n, that is $n = 3, 4$, the inequality sign in the formula (A) can be omitted. Further, we shall show in the sequel that, at least for the values up to $n = 12$ inclusive, and so for all practical values of n, our problem will indeed be solved exhaustively by the *equation*

(B)
$$p_n = \frac{(n-3)(n-4)+2\alpha_n}{12}.$$

For the following values of p, namely $p = p_n, p_n + 1, p_n + 2, \ldots, p_{n+1} - 1$, where n is such a value for which equation (B) is valid, the converse question of the maximum number n_p of neighbouring elements on a surface of genus p will then be answered by the equation

(C)
$$n_p = \frac{7}{2} + \frac{1}{2}\sqrt{1+48p-8\alpha_p},$$

where α_p denotes the smallest positive integer which makes the radicand into a square. (In this form, with the addition of the sign $<$, the result appears in Heawood [7A], if there the value κ, of which Heawood only says it is 'a positive constant depending on the connectivity of the complete surface', be put equal to $2p-2$.) Hence if $\alpha_p < 6$, then the value of p employed would be such that the corresponding number of neighbouring regions n_p did not already occur on a surface of smaller genus. If $\alpha_p \geq 6$, namely $= 6\nu + \alpha_p'$, where $\alpha_p' < 6$, then the same number of neighbouring regions already occurs on a surface of genus $p-\nu$.

Before we begin the promised proof, the following remarks necessary for understanding it must be stated.

9. The numbers α have a simple significance with reference to the nature of the network on the surface. If there are n points on any surface of genus p, and the resulting faces are all simply-connected, then the relation

$$p = \frac{(n-3)(n-4)+2\alpha}{12}$$

certainly always holds, whence the number of faces is $\dfrac{n(n-1)-\alpha}{3}$. Thus in the whole network there are now exactly $n(n-1)$ polygonal vertices or angles, so α is the number of polygonal vertices which remain if three vertices are removed from each face. Hence if equation (B) holds, the faces corresponding to the four possible values of α_n (0, 2, 3, 5) form respectively:

1) triangles;

or

2) triangles and 2 quadrangles
   or triangles and 1 pentagon;

or

3) triangles and 3 quadrangles
   or triangles, 1 quadrangle, 1 pentagon
   or triangles and 1 hexagon;

or

4) triangles and 5 quadrangles
   or. . .
   or finally: triangles and 1 octagon.

Of particular interest is the case where $\alpha_n = 0$, that is, supposing that equation (B) holds, when the whole surface is divided into $\dfrac{n(n-1)}{3}$ triangles, of which every $n-1$ meet in a vertex; or, in the dual interpretation, in n $(n-1)$-gons of which every three form a polyhedral vertex. We shall call these cases *regular*. They arise if $\alpha_n = 0$, so that n has one of the four forms:

$$n = 12\kappa,$$
$$n = 12\kappa + 3,$$
$$n = 12\kappa + 4,$$
$$n = 12\kappa + 7,$$

because the genus p of the surface, of which such regular division is possible, satisfies the *necessary* condition

$$1 + 48p = \lambda^2$$

or
$$p = \frac{\lambda^2 - 1}{48},$$

where
$$\lambda^2 \equiv 1 \ \text{mod.} \ 48.$$

Put $\kappa = 0$; then from the third line one has $n = 4$, and consequently $p = 0$ (such a regular case is for example the surface of the tetrahedron). Indeed, all regular cases are formed by the generalizations of the tetrahedron to surfaces of higher genus in the sense of analysis situs, and of course in one or other of their two dual conceptions; just as we consider the tetrahedron (for which the dual manifestations coincide) as a system of four faces each of which borders the three others, or as a network of four points, each of which is connected to the three others on the surface.

§3

Arithmetical form of the investigation.—Proof of the equation (B) for $n = 4, 5, \ldots, 12$.

10. The difficulties inherent in the attempt to prove equation (B) have their origin in the purely arithmetical nature of the problem. Indeed a purely arithmetical form can be given to it. One considers n $(n-1)$-gons (dually, n simply-connected faces as the carriers of n points, from each of which $n-1$ lines go across the boundary), labels them by writing thereon the numbers $1, 2, \ldots, n$, and then labels each of the sides of the polygon κ by one of the numbers $1, 2, \ldots, \kappa - 1, \kappa + 1, \ldots, n$, in order to indicate to which other polygon the polygon κ is to be joined along this edge. Then, if one fastens them together in the manner indicated, a closed surface is always obtained which is divided into n regions; but the genus of it will in general exceed $\dfrac{(n-3)(n-4) + 2\alpha_n}{12}$. If this number is achieved, then the total number of vertices is $\dfrac{n(n-1) - \alpha_n}{3}$.

The arrangement of λ polygons, $1, 2, \ldots, \lambda$ for example, meeting at one vertex (or λ points $1, 2, \ldots, \lambda$ being vertices of one polygon), is thence expressed arithmetically by the sequence of them in a clockwise order around the vertex. If, for instance, the order is the natural one $1, 2, \ldots, \lambda$, then among the $n-1$ numbers which label the sides of 1 are: $2, \lambda$ (following in that order); and in polygon 2 we have: $3, 1$; for $3: 4, 2$; and so on. We shall say that these λ number pairs form a *cycle of λ elements or pairs*. In particular, therefore, a trivalent vertex (or triangle) gives a three element cycle

$$(i\kappa), \quad (\kappa l), \quad (li);$$

and then it follows that among the boundary segments of the polygon i, and in clockwise succession, are κ, l; in κ are l, i; and in l are i, κ.

So our problem is now this (where the numbers in each line are to be considered as closed into a cycle):

(D) $\begin{cases} (1) \ \text{the numbers } 2, 3, \ldots, n \ \text{(which represent the sides of polygon 1),} \\ (2) \ \text{the numbers } 1, 3, \ldots, n \ \text{(which represent the sides of polygon 2),} \\ \qquad \qquad \cdot \quad \cdot \quad \cdot \quad \cdot \quad \cdot \\ (n) \ \text{the numbers } 1, 2, \ldots, n-1 \ \text{(which represent the sides of polygon } n), \end{cases}$

are to be arranged so that the number of cycles arising is exactly $\dfrac{n(n-1)-\alpha_n}{3}$. For the sake of brevity we shall denote such an array by (D).

11.   For each of the numbers $n = 4, 5, 6, \ldots, 11, 12$, the following tables give an arrangement (D) by which the validity of statement (B), and simultaneously (C), is proved. Above each table is placed the value of n and of p_n, α_n, and $F = \dfrac{n(n-1)-\alpha_n}{3}$. In the tables the number pairs which belong to cycles of more than three elements are marked by a curve placed under them, while under each table we give the number of triangles, and those cycles which have more than three elements.

$n = 4$, $p_4 = 0$, $\alpha_4 = 0$, $F = 4$.
Regular (tetrahedron).
1) 2 3 4
2) 3 1 4
3) 4 1 2
4) 2 1 3
4 triangles.

$n = 5$, $p_5 = 1$, $\alpha_5 = 5$, $F = 5$.
1) 3 2 4 5
2) 4 3 5 1
3) 5 4 1 2
4) 1 5 2 3
5) 2 1 3 4
5 quadrangles: [1342], [1435], [1254], [1523], [3245].

In the table for $n = 5$ a curve should strictly be placed under every pair, and they are all omitted for that reason; the arrangement is illustrated in Fig. [7.8]. The closed surface of genus one is obtained by joining together the lines AB and DC, BC and AD. The result is a 'pentahedron', for which, just as for the tetrahedron, the two dual manifestations coincide. The five vertices of the pentahedron are neighbouring points. (Thus we have another surface, regularly divided into five quadrangles, and with genus greater than zero, which is not contained in the four series of regular cases cited above. One can easily convince oneself that, in our problem where the least genus is always sought, other regular divisions cannot arise. Letting that restriction fall, the tetrahedron and pentahedron are just the first cases in an infinite series of regular surfaces, to which I hope to return when the opportunity arises. These are surfaces of genus $\dfrac{(n-1)(n-4)}{4}$ ($n = 4, 5, 8,$ $9, \ldots, 4\kappa, 4\kappa + 1, \ldots$), which are divided into n neighbouring $(n-1)$-gons so that

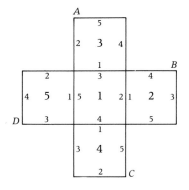

Fig. 7.8.

just n vertices arise, which are simultaneously neighbouring points. Compare the previous figure and the following Figure [7.9], also the figures of W. Dyck [9].)

$n = 6$, $p_6 = 1$, $\alpha_6 = 3$, $F = 9$.

1) 3 5 2 6 4
2) 4 6 3 1 5
3) 5 1 4 2 6
4) 6 2 5 3 1
5) 1 3 6 4 2
6) 2 4 1 5 3

8 triangles and the hexagon
[123456].

$n = 7$, $p_7 = 1$, $\alpha_7 = 0$, $F = 14$.
Regular.

1) 2 4 3 7 5 6
2) 3 5 4 1 6 7
3) 4 6 5 2 7 1
4) 5 7 6 3 1 2
5) 6 1 7 4 2 3
6) 7 2 1 5 3 4
7) 1 3 2 6 4 5
14 triangles.

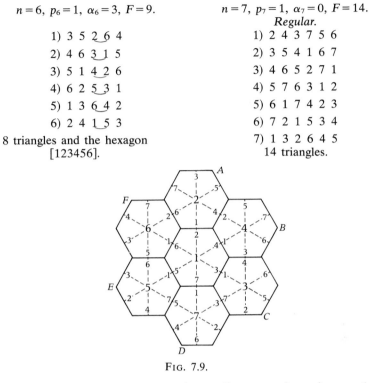

FIG. 7.9.

Fig. [7.9] illustrates the arrangement for $n = 7$ on a surface of genus 1. The continuous lines divide the surface into seven neighbouring hexagons; the dotted lines join the middle points of these with one another. The closed surface results by joining to one another the lines AB and ED, BC and FE, CD and AF.

$n = 8$, $p_8 = 2$, $\alpha_8 = 2$, $F = 18$.

1) 2 3 4 5 6 7 8
2) 3 1 8 5 4 7 6
3) 4 1 2 6 7 5 8
4) 2 1 3 8 6 5 7
5) 6 1 2 8 3 7 4
6) 2 1 5 4 8 7 3
7) 8 1 2 4 5 3 6
8) 2 1 7 6 4 3 5

16 triangles and the two quadrangles [1425], [1627].

*   *   *   *   *

Heffter's tables for $n = 5$, 6, and 7 have the property that all rows may be obtained from the first row by cyclic permutations. In the remainder of his paper, after presenting tables for $n = 9$, 10, 11, and 12, he went on to show that such a cyclic arrangement is possible only when $n = $ 5, 6, or n is of the form $12\kappa + 7$. He then gave an explicit construction for this case, under the additional restrictions that $4\kappa + 3$ is a prime number ($ = q$, say) and the order of 2 in the multiplicative group of residues modulo q is either $q - 1$ or $\frac{1}{2}(q - 1)$. These conditions define a sequence which begins $n = 19$, 31, 55, 67, 139, 175, 199, ..., but even today we do not know whether or not the sequence continues indefinitely. A leading twentieth-century mathematician, Emil Artin, has made a deep conjecture in number theory, whose truth would imply that Heffter's sequence is infinite; unfortunately Artin's conjecture has been proved only by assuming the truth of an even more enigmatic conjecture—the 'generalized Riemann hypothesis' [10]. So Heffter never knew if he had proved $\chi(S_p) = h(p)$ for infinitely many values of p. In fact, the question is now an academic one, for it has been shown (in a similar, but independent, way) that $\chi(S_p) = h(p)$ for all $p \geqslant 1$ [11]. Although Heawood and Heffter each lived to a ripe old age—Heawood died in 1955 at the age of 93, and Heffter in 1962 at 99—neither of them lived to see the complete solution of the problem in 1967.

One-sided surfaces

A good account of the strange properties of one-sided surfaces may be found in a little book by Fréchet and Fan [12]. We shall give a brief, self-contained outline, referring the reader to that book for more details.

Let us begin with an illustration: an insect crawling on the inside surface of a sphere is trapped, for it cannot find any route which will bring it onto the outside surface. The inside and outside surfaces are quite distinct, and so we say that the sphere is a two-sided surface. This property of two-sidedness is shared by the torus (S_1) and all the closed surfaces (S_p) constructed in Chapter 5 by attaching handles to a sphere. However, in the second half of the nineteenth century, mathematicians realised that two-sidedness is not an inherent property of all surfaces. An example of a surface having only one side was discovered, in about 1858, by Listing [13] and Möbius [14] independently (see also [15], [16]); this surface is now known as the *Möbius strip*.

The construction of a Möbius strip is quite simple. We take a rectangular strip $ABCD$, give it a half-twist, and join the ends AD, BC so that A coincides with C and B with D (Fig. 7.10). If our wandering insect starts at a point Q on the 'equator' and crawls once around the strip, it will find itself back at the same point Q, but on the 'other side' of the surface. There is, in fact, only one side!

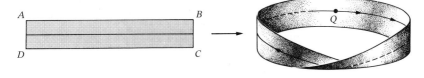

FIG. 7.10.

It is natural to ask if one-sidedness can occur in a closed surface—the Möbius strip is not closed, for it has a boundary curve. Such surfaces do exist, but in order to construct them we must renounce the limitations of three-dimensional space. This is no hindrance to the mathematician, for whom spaces of many dimensions are a commonplace, but it does mean that the visualization of these surfaces is rather complicated.

First, let us consider the boundary curve of a Möbius strip. It is formed from the two edges AB, CD of a rectangle by joining A to C and B to D, and so, topologically, it is just a circle; certainly, the circle is twisted and looped, but nevertheless it is still a circle. We now imagine that the boundary circle of a Möbius strip is untwisted and unlooped, so that it becomes a flat circle lying in a plane. Although this is mathematically possible, it cannot be done in ordinary three-dimensional space without causing the strip to intersect itself. An attempt to picture the result of flattening the boundary circle in this way is shown in Fig. 7.11; a Möbius strip in this form is called a *cross-cap*. In order to construct a closed one-sided surface, we take a sphere, cut out a piece with a circular boundary, and attach a cross-cap in its place. Clearly, we could be more general: if q circular holes are made and q cross-caps attached thereto, then we still get a closed one-sided surface. So we have an infinite family $\Sigma_1, \Sigma_2, \ldots$ of surfaces which are both closed and one-sided. Indeed, it

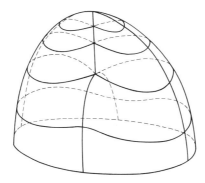

FIG. 7.11.

may be shown that any surface which has these two properties is topologically equivalent to one of the surfaces Σ_q [12]. Just as for the two-sided surfaces, the Euler characteristic

$$\eta(M) = N_v - N_e + N_f$$

is the same for all maps M on Σ_q, and its constant value turns out to be

$$\eta(\Sigma_q) = 2 - q.$$

The colouring of maps on one-sided surfaces was investigated by Heinrich TIETZE at the beginning of this century, using Heffter's formulation of the problem in terms of neighbouring regions. He showed [7C] that the Möbius strip admits six neighbouring regions and deduced that its chromatic number is six. For the closed surface Σ_1 he found $\chi(\Sigma_1) = 6$, but the chromatic number of Σ_2 eluded him. The latter surface, which Tietze called the 'closed two-fold one-sided surface', is now usually known as the *Klein bottle*—it is a somewhat inefficient bottle, since it has no inside, but the reason for its name may be apparent from the sketch in Fig. 7.12.

Tietze also applied the arguments of Heawood to obtain an upper bound for $\chi(\Sigma_q)$. His result may be derived by substituting Σ_q for S and $2 - q$ for $\eta(S)$ in the general formula given earlier in this chapter. We find, for $q \geqslant 2$,

$$\chi(\Sigma_q) \leqslant h(\Sigma_q) = [\tfrac{7}{2} + \tfrac{1}{2}\sqrt{(1 + 24q)}].$$

A translation of part of Tietze's paper follows. About half of the original consisted of footnotes dealing with historical matters (mentioned elsewhere in this book) and terminology; we have suppressed most of the footnotes in the translation.

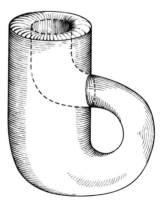

FIG. 7.12.

7C

H. TIETZE

EINIGE BEMERKUNGEN ÜBER DAS PROBLEM DES KARTENFÄRBENS AUF
EINSEITIGEN FLÄCHEN

[Some remarks on the problem of map-colouring on one-sided surfaces]

Jahresbericht der Deutschen Mathematiker-Vereinigung **19** (1910), 155–159.

That five neighbouring regions (spatia confinia) can be constructed on a Möbius strip [14, p. 482], that is five regions each of which touches the others along at least one edge, was reported by L. Heffter in his work 'Über das Problem der Nachbargebiete' [7B]. As will be shown here, it is possible for six, but no more, neighbouring regions to be marked on a Möbius strip.

1. If the edge AB of the rectangle in the figure were to be brought into coincidence with the edge $A'B'$ (and, therefore, A with A', B with B') then the arrangement of regions illustrated would furnish an example of six neighbouring regions on the resulting Möbius strip, provided that we coalesce into one region on the strip regions labelled with the same numbers (e.g. 5) but separated in the figure. So at least six neighbouring regions can be constructed on any one-sided surface, since on any one-sided surface there are areas which represent a Möbius strip.

2. The simple adaptation of an argument employed by Heawood [7A] for two-sided surfaces to the case of one-sided surfaces now shows that not more than six neighbouring regions can exist on the Möbius strip. Henceforth let N be the value of the Euler characteristic $\alpha_0 - \alpha_1 - \alpha_2$, attached to a given surface $\mathscr{F}$ where α_0, α_1, α_2 denote the number of vertices, edges, faces of such a division. For an arbitrary map drawn on $\mathscr{F}$ (that is, for an arbitrary division of $\mathscr{F}$ into bounded regions which need not necessarily be faces), we obtain the equation

$$(1) \qquad e - k + \sum_{i=1}^{i=f} N_i = e - k + f + \sum_{i=1}^{i=f} (N_i - 1) = N,$$

where e, k, f are the number of vertices, edges and components of this division and N_i is the value of N pertaining to the ith component. Moreover, denoting by k_h the number of edge-ends which are incident with the hth vertex, then by virtue of

$$2k = \sum_{h=1}^{h=e} k_h,$$

we obtain from (1) the relation

(2)
$$k = 3\left[f - N + \sum_{i=1}^{i=f} (N_i - 1)\right] - \sum_{h=1}^{h=e} (k_h - 3).$$

Since for all components, whether two- or one-sided, we have $N_i \leq 1$, and assuming throughout the following investigation that all $k_h \geq 3$, it follows that

(3)
$$A_f = \frac{2k}{f} \leq 6\left(1 - \frac{N}{f}\right),$$

where A_f is the average number of edges which are contained in a face. For the simplest type of closed one-sided surface, as represented for example by the projective plane, $N = 1$, so that $A_f < 6$. Consequently, for each division of such a surface, at least one region is contiguous with at most 5 other regions. Thus it follows by a simple induction proof (see [7A]) that, for *every* map on the surface, 6 colours certainly suffice to give each of the regions one of the colours, in such a way that no two regions which are adjacent to one another along one or more edges exhibit the same colour.

In general, the smallest number c of colours which suffice to colour each map on a surface in the specified manner, is never smaller than the largest possible number v of neighbouring regions on the surface. Thus, for every one-sided surface which can be regarded as a portion of the closed one-sided surface under consideration, and also for this surface itself, the equation:

$$c = v = 6$$

follows.

Thus, in particular, *for all maps on the projective plane or the Möbius strip, the regions can be distinguished by 6 colours, and so not more than 6 neighbouring regions may be drawn.*

3.   For the closed 2-fold one-sided surface, $N = 0$ and $A_f \leq 6$, implying that $c \leq 7$; whilst on the other hand, just as for every one-sided surface, $v \geq 6$. Whether there is a map on this surface whose colouring actually requires 7 colours, and in particular whether it has 7 neighbouring regions, remains undecided. In any case, 7 such neighbouring regions, in as much as they fill up the whole closed surface, must be the totality of faces; this follows from (2) (under the permissible assumption $k_h \geq 3$) since $k \geq \dfrac{7.6}{2}$, or by considerations analogous to those employed by Heffter [7B].

If $q > 2$, then for a closed q-fold one-sided surface we have $N < 0$. So from (3), as Heawood has shown for c, and hence also for v,

$$v \leq c \leq E\left(\frac{7 - \sqrt{(49 - 24N)}}{2}\right) = E\left(\frac{7 + \sqrt{(1 + 24q)}}{2}\right)$$

where $E(x)$ denotes the largest integer $\leq x$.

4.   For several years I have given a lecture [17] concerning the facts mentioned in Sections 1 and 2, and on those occasions I have also exhibited the 6 regions on the Möbius strip using a model made of coloured materials. In the meantime, I have noticed that the presence of 6 neighbouring regions on the one-fold one-sided surface is already contained implicitly in the first of the examples of one-sided surfaces given by Möbius [14]. It is a polyhedral surface ([18] p. 39 and figure) which in a concrete realization can be obtained in the following manner.

Let A, B, C, D, E be the points with rectangular coordinates $(0, 0, 0)$, $(1, 0, 0)$, $(0, 1, 0)$, $(\frac{2}{3}, \frac{2}{3}, \frac{2}{3})$, $(0, 0, 1)$, and let F be any point which does not lie in a plane with any of the 5 triangles (d) (see below). The said polyhedron consists of the 10 triangles:

$$(d) \quad ABC, \quad BCD, \quad CDE, \quad DEA, \quad EAB$$
$$(d') \quad FAC, \quad FBD, \quad FCE, \quad FDA, \quad FEB.$$

It is a closed one-fold one-sided surface of the same type as the projective plane. Each of the 6 vertices is connected with every other by an edge passing through the surface. By transition to the reciprocal (dual) figure of the surface [7B], we obtain 6 regions, of which each one borders every other along one edge of the region; that is, 6 neighbouring regions.

It is worth emphasizing that Tietze was unable to obtain the chromatic number of the Klein bottle, Σ_2. As he pointed out, the existence of six neighbouring regions on a Möbius strip, and hence on any one-sided closed surface, means that $\chi(\Sigma_2) \geqslant 6$; the general 'Heawood bound' gives $\chi(\Sigma_2) \leqslant 7$.

In the 1930s these questions were taken up by several American mathematicians, as part of a resurgence of interest in colouring problems. In 1934 Philip FRANKLIN [19] proved that $\chi(\Sigma_2) = 6$, a rather surprising result, since it means that the Heawood bound is not attained in this case. However, I. N. Kagno [20] then showed that if $q = 3$, 4, and 6, then $\chi(\Sigma_q) = h(\Sigma_q)$. We now know the complete solution of the map colouring problem for one-sided surfaces [21], and it turns out that the Klein bottle is truly anomalous. It is the only closed surface S for which $\chi(S)$ is not equal to $h(S)$—with the additional tantalizing possibility that this may be so for the sphere.

References

1. BALTZER, R. Eine Erinnerung an Möbius und seinen Freund Weiske. *Ber. K. Sächs. Ges. Wiss. Leipzig Math.-Phys. Cl.* **37** (1885), 1–6.
2. MADDISON, I. Note on the history of the map-coloring problem. *Bull. Amer. Math. Soc.* **3** (1896–7), 257.
3. ROUSE BALL, W. W. *Mathematical recreations and essays* (11th edn.). Macmillan, London, 1939.
4. BELL, E. T. *The development of mathematics* (2nd edn.). McGraw-Hill, New York, 1945.
5. COXETER, H. S. M. The four-color map problem. *Math. Teacher* **52** (1959), 283–289.
6. MAXWELL, J. C. On reciprocal figures and diagrams of forces. *Phil. Mag.* (4) **27** (1864), 250–261 = *Sci. Papers*, Vol. 1, 514–525.
7. MAXWELL, J. C. On reciprocal figures, frames and diagrams of forces. *Trans. Roy. Soc. Edinb.* **26** (1869–72), 1–40 = *Sci. Papers*, Vol. 2, 161–207.
8. HEATH, T. L. *The thirteen books of Euclid's elements*, Vol. 3. Dover, New York, 1956.
9. DYCK, W. Gruppentheoretische Studien. *Math. Ann.* **20** (1882), 1–44.
10. HOOLEY, C. On Artin's conjecture. *J. Reine Angew. Math.* **225** (1967), 209–220.
11. RINGEL, G. and YOUNGS, J. W. T. Solution of the Heawood map-coloring problem. *Proc. Nat. Acad. Sci. U.S.A.* **60** (1968), 438–445.

12. FRÉCHET, M. and FAN, K. *Initiation to combinatorial topology*. Prindle, Weber and Schmidt, Boston, 1967.

13. LISTING, J. B. Der Census räumlicher Complexe oder Verallgemeinerung des Euler'schen Satzes von den Polyëdern. *Abh. K. Ges. Wiss. Göttingen Math. Cl.* **10** (1861–2), 97–182 = Separately as book, Dieterich'sche Verlagshandlung, Göttingen, 1862.

14. MÖBIUS, A. F. Über die Bestimmung des Inhaltes eines Polyëders. *Ber. K. Sächs. Ges. Wiss. Leipzig Math.-Phys. Cl.* **17** (1865), 31–68 = *Werke*, Vol. 2, 473–512.

15. STÄCKEL, P. Die Entdeckung der einseitigen Flächen. *Math. Ann.* **52** (1899), 598–600.

16. MÖBIUS, A. F. Zur Theorie der Polyëder und der Elementarverwandtschaft (Mittheilungen aus Möbius' Nachlass, ed. C. Reinhardt). *Werke*, Vol. 2, 515–559.

17. [Meeting of the Mathematical Society of Vienna, 10 February 1905]. *Jahresber. Deut. Math.-Ver.* **14** (1905), 202.

18. KLEIN, F. *Elementar-Mathematik vom höheren Standpunkte aus*, Vol. 2. Teubner, Leipzig, 1911. English edn.: *Elementary mathematics from an advanced standpoint, geometry*. Dover, New York, 1939.

19. FRANKLIN, P. A six-color problem. *J. Math. Phys.* **13** (1934), 363–369.

20. KAGNO, I. N. A note on the Heawood color formula. *J. Math. Phys.* **14** (1935), 228–231.

21. RINGEL, G. *Färbungsprobleme auf Flächen und Graphen*. Deutscher Verlag der Wiss., Berlin, 1959.

8

Ideas from Algebra and Topology

G. R. KIRCHHOFF (1824–87)

THE branches of mathematics are not isolated from one another, but are interlinked in many ways; for instance, in Chapters 1 and 2 we described how the origins of graph theory are related to those of topology. One of the aims of this chapter is to show how the growth of topology, in the years from 1860 to 1930, led to a second infusion of topological ideas into the study of graphs.

Another important factor in the development of both topology and graph theory was the use of techniques from modern algebra. The first such example of the use of algebraic methods was due to a physicist, rather than to a mathematician, and we shall begin by explaining the relevant background, which is concerned with the structure of the set of circuits in a graph.

The algebra of circuits

In 1845 a young student of physics, Gustav Robert KIRCHHOFF, formulated two rules which govern the flow of electricity in a network of wires [1]; these rules are now known as Kirchhoff's laws. From our point of view, such a network may be thought of as a graph, with certain physical characteristics which need not concern us here. One of Kirchhoff's laws deals with the current flow around a circuit, and it leads to a set of linear equations, one for each circuit in the graph. These equations are not all independent, and so there arises the question of how many of them are necessary in order to obtain the complete set. It was this question which Kirchhoff asked and answered in 1847 [8A].

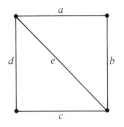

F𝐼G. 8.1.

Let us consider the graph-theoretical implications of this problem, by means of a simple example. The graph shown in Fig. 8.1 contains three circuits, which we may denote by their sets of edges: $C_1 = abe$, $C_2 = cde$, and $C_3 = abcd$. It is clear that the circuit C_3 is, in some sense, the 'sum' of C_1 and C_2. To make this precise, we shall say that the **sum** of two circuits consists of all those edges which belong to one, but not to both, of them; this can be extended in an obvious way to the sum of any finite number of circuits. A set of circuits is said to be **independent** if no one of them can be expressed as a sum of others, so that in our example, the sets $\{C_1, C_2\}$ and $\{C_1, C_3\}$ are independent, whereas the set $\{C_1, C_2, C_3\}$ is not. A maximal set of independent circuits is called a **fundamental set.**

Our problem is to find the number of circuits in a fundamental set, and to describe a method for constructing such a set; if a fundamental set is given, then every other circuit can be written as a sum of members of this set. The problem treated by Kirchhoff is essentially the same—the linear equation corresponding to a sum of circuits is just the sum of the equations corresponding to the individual circuits, and consequently the equations corresponding to an independent set of circuits are independent.

In his 1847 paper [8A], Kirchhoff explained how a fundamental set of circuits may be constructed (Section 1), and proved that, for any connected graph with m vertices and n edges, a fundamental set always contains $n - m + 1$ circuits (Section 5). We have omitted those parts of his paper in which he dealt with the technicalities of determining the currents in the wires of the network, but the statement of the result of this calculation has been retained.

8A

G. R. KIRCHHOFF

ÜBER DIE AUSFLÖSUNG DER GLEICHUNGEN, AUF WELCHE MAN
BEI DER UNTERSUCHUNG DER LINEAREN VERTHEILUNG
GALVANISCHER STRÖME GEFÜHRT WIRD

[On the solution of the equations obtained from the investigation of the linear
distribution of galvanic currents]

Annalen der Physik und Chemie **72** (1847), 497–508.

If we are given a system of n wires, $1, 2 \ldots n$, which are joined to one another in an arbitrary way, and if the system has an arbitrary electromotive force in series with each of the wires, then the number of equations necessary for determining the strengths of the currents $I_1, I_2 \ldots I_n$, flowing through the wires, is obtained by applying the following two theorems [1]:

I. If the wires $k_1, k_2, \ldots$ form a closed figure, and if w_k denotes the resistance of the wire k, and E_k denotes the electromotive force in series with it, taken to be positive in the same direction as I_k, then in the case when $I_{k_1}, I_{k_2}, \ldots$ are all considered as positive in the *same* direction:

$$w_{k_1}I_{k_1} + w_{k_2}I_{k_2} + \ldots = E_{k_1} + E_{k_2} + \ldots.$$

II. If the wires $\lambda_1, \lambda_2, \ldots$ meet together in one point, and if $I_{\lambda_1}, I_{\lambda_2}, \ldots$ are all taken to be positive towards this point, then:

$$I_{\lambda_1} + I_{\lambda_2} + \ldots = 0.$$

Assuming that the given system of wires does not decompose into quite separate parts, I shall now prove that the solutions of the equations, which are obtained for $I_1, I_2, \ldots, I_n$ by using these theorems, can be stated in general as follows:

Let m be the number of available crossing-points, i.e. the points where two or more wires meet, and let $\mu = n - m + 1$. Then the common denominator of all the currents I is the sum of those combinations of μ elements $w_{k_1} . w_{k_2} . \ldots . w_{k_\mu}$, chosen from $w_1, w_2, \ldots w_n$, which have the property that, after the removal of the wires $k_1, k_2, \ldots k_\mu$, no closed figure remains. The numerator of I_λ is the sum of those combinations of $\mu - 1$ elements, $w_{k_1} . w_{k_2} . \ldots . w_{k_{\mu-1}}$, chosen from $w_1, w_2 \ldots w_n$, which have the property that, after the removal of $k_1, k_2, \ldots k_{\mu-1}$, *one* closed figure remains, and contains λ. Each combination is multiplied by the sum of the electromotive forces occurring in the relevant closed figure. The electromotive forces are taken to be positive in the direction in which I_λ is taken to be positive.

For the sake of simplicity, I shall divide my proof of this theorem into separate sections.

1.

Let μ be the *least* number of wires that must be removed from an arbitrary system so that all the closed figures are destroyed; then μ is also the number of independent equations which can be obtained by using Theorem I.

For, one can find μ equations which are independent of one another, and from which every equation obtained from Theorem I may be derived, as follows:

Let $1, 2 \ldots \mu - 1, \mu$ be μ wires whose removal leaves no closed figure, and such that the removal of any $\mu - 1$ of these wires leaves *one* closed figure. Let us apply Theorem I to the closed figures which remain if

$$2, 3, \ldots \mu$$
$$1, \quad 3, \ldots \mu$$
$$\cdots \cdots \cdots \cdots$$
$$1, 2, 3, \ldots \mu - 1$$

respectively, are removed.

None of the μ equations formed in this way can be a consequence of the others, since each one contains an unknown which does not appear in any others; only the first contains I_1, the second I_2, and so on. But from these equations, every other equation which can be obtained from Theorem I can be derived; for an equation derived from a closed figure made up from several other closed figures, must itself be obtainable from the equations for these other figures (by addition and subtraction); and, as we wish to show, every closed figure can be made up from those μ figures. The collection of closed figures of the given system, which we shall denote by S, can be divided into those containing the wire μ, and those contained in the system S' obtained from S by removing the wire μ. We suppose that all of the figures of the second type can be made up from the first $\mu - 1$ of those μ figures, and so we see that every figure of the system S may be made up from all μ of them. For, an arbitrary figure containing the wire μ is made up from a particular one containing μ, and others to which μ does not belong. But the assumption made about the system S' may be reduced to a similar assumption with regard to S'', where S'' is the system which arises from S by removing both μ and $\mu - 1$; namely to the assumption that all closed figures occuring in S'' can be made up from the first $\mu - 2$ of those μ. By continuing in this way, we eventually arrive at the system $S^{(\mu-1)}$; since this contains only one closed figure, the correctness of the assumption we must make for this system in order to see the truth of our assertion, is itself clear.

$$* \quad * \quad * \quad * \quad *$$

5.

In order to prove our theorem, as we stated it, we must still show that $\mu = n - m + 1$. This statement holds only if the given system of wires does not decompose into quite separate parts, whereas our discussion so far did not require such a hypothesis.

As we have seen, μ is the number of independent equations which can be obtained from Theorem I; therefore the number of independent equations obtained from Theorem II must be $n - \mu$. But it will now be shown that, under our hypothesis, this number is $m - 1$, whence it follows that $\mu = n - m + 1$. One cannot obtain more than $m - 1$ independent equations from Theorem II; if we apply that theorem to all m of the crossing-points, then each current I appears twice in the resulting equations, once with the coefficient $+1$, the other time with the coefficient -1; the sum of all these equations thus gives the identity $0 = 0$. But the equations obtained by applying that theorem to $m - 1$ arbitrary crossing-points are independent of one another, since they have the property that if we choose any number of them, at least one of the unknowns appears only once.

Let us call the crossing-points $1, 2, \ldots m$, and let us call (κ, λ) a wire joining the 2 points κ and λ. Then if, in the equations corresponding to the points $\kappa_1, \kappa_2 \ldots \kappa_\nu$, one of those points ($\kappa_1$, say) is joined to another point λ (other than $\kappa_2 \ldots \kappa_\nu$), the unknown $I_{(\kappa_1, \lambda)}$ will occur only once. But if the wires joining the points $\kappa_1, \kappa_2 .. \kappa_\nu$ to one another do not form a self-closed system, then one of the points $\kappa_1, \kappa_2 .. \kappa_\nu$ must also be joined to a point λ, as well as being joined to others of the same set.

*   *   *   *   *

Kirchhoff's laws soon found their rightful place among the basic facts of elementary physics. But the mathematical techniques underlying his general method of solution were of less immediate practical value, and their importance was not recognized until mathematicians began to employ algebraic methods in topology.

The two classic works of Listing, the *Topologie* and the *Census*, heralded the investigation and application of topological ideas in several branches of mathematics, a process which developed sporadically throughout the second half of the nineteenth century. Then, in the years 1895–1904, a series of papers by Henri Poincaré laid the foundations of the subject now known as 'algebraic topology'.

Among other things, Poincaré developed [2] a method for constructing geometrical objects, which he called 'complexes' (following Listing), from basic building blocks, called 'cells'. In order to describe how the cells fit together, he adapted Kirchhoff's technique, replacing the system of linear equations by a matrix. The simplest kinds of cells are the 0-cells (vertices) and the 1-cells (edges); a complex constructed from such cells is a graph, and the matrix which Poincaré used to describe how the cells are fitted together in this case is now called the 'incidence matrix' of the graph. All of this will be clarified presently.

Poincaré's theory was an instant success. One of the first books to include an account of it was the third volume of the famous *Encyklopädie der Mathematischen Wissenschaften*, a monumental work which was intended to survey all the mathematical knowledge available at the time. Its twenty-three volumes, published from 1898 to 1935, contained articles by eminent scholars on all branches of mathematics. The article on topology (Analysis Situs) was written in 1907 by M. Dehn and P. Heegaard [3] and is of historical importance in graph theory, since it began with a section on graphs (called *Liniensysteme*). This was the first comprehensive account of the subject, and it contained an extensive set of references, not all of them accurate.

At about this time topological ideas began to spread in the United States of America, mainly through the influence of George David BIRKHOFF and Oswald VEBLEN. (In addition to their general interest in topology, both men were intrigued by the four-colour conjecture, and in the next chapter we shall give an account of their work on that notorious problem.) In 1916, Veblen gave a series of American Mathematical

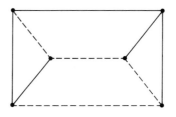

F<small>IG</small>. 8.2.

Society Colloquium lectures on Analysis Situs, and these were later published as a book. Like the authors of the *Encyklopädie* article, Veblen chose to introduce the general theory by means of a chapter on graphs.

In the extract [8B] which we give here, Veblen showed clearly how the ideas of Kirchhoff had been formalized by Poincaré. One central concept is the number of circuits in a fundamental set. This number had already appeared in various guises—as the 'cyclomatic number' in Listing's *Census*, and the 'degree of continuity' (less one) in Jordan's work on trees [3B]. In the algebraic treatment it turns out to be the dimension of the kernel of the incidence matrix, and so it is related to the rank of this matrix.

Veblen also discussed Kirchhoff's method for constructing a fundamental set of circuits, in a form which clarified its relationship with the problem of finding a special kind of tree in a graph (Sections 26 and 27 of [8B]). He showed that every connected graph contains a tree which includes every vertex of the graph; in modern terminology, this is called a **spanning tree.** For example, a spanning tree in the graph depicted in Fig. 8.2 is indicated by the unbroken edges. From each spanning tree in a graph it is possible to construct a fundamental set of circuits, in the manner explained by Veblen at the end of our extract.

8B

O. VEBLEN
LINEAR GRAPHS

Analysis Situs. (American Mathematical Society Colloquium Lectures 1916), New York, 1922.

*   *   *   *   *

Symbols for Sets of Cells

14. Let us denote the 0-cells of a one-dimensional complex C_1 by a_1^0, a_2^0, ..., $a_{\alpha_0}^0$ and the 1-cells by a_1^1, a_2^1, ..., $a_{\alpha_1}^1$.

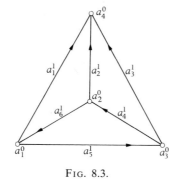

$$\text{FIG. 8.3.}$$

Any set of 0-cells of C_1 may be denoted by a symbol $(x_1, x_2, \ldots, x_{\alpha_0})$ in which $x_i = 1$ if a_i^0 is in the set and $x_i = 0$ if a_i^0 is not in the set. Thus, for example, the pair of points a_1^0, a_4^0 in Fig. [8.3] is denoted by $(1, 0, 0, 1)$. The total number of symbols $(x_1, x_2, \ldots, x_{\alpha_0})$ is 2^{α_0}. Hence the total number of sets of 0-cells, barring the null-set, is $2^{\alpha_0} - 1$. The symbol for a null-set, $(0, 0, \ldots, 0)$ will be referred to as *zero* and denoted by 0.

The marks 0 and 1 which appear in the symbols just defined, may profitably be regarded as residues, modulo 2, i.e., as symbols which may be combined algebraically according to the rules

$$0+0=1+1=0, \qquad 0+1=1+0=1,$$
$$0\times 0=0\times 1=1\times 0=0, \qquad 1\times 1=1.$$

Under this convention the *sum* (mod. 2) of two symbols, or of the two sets of points which correspond to the symbols $(x_1, x_2, \ldots, x_{\alpha_0}) = X$ and $(y_1, y_2, \ldots, y_{\alpha_0}) = Y$, may be defined as $(x_1 + y_1, x_2 + y_2, \ldots, x_{\alpha_0} + y_{\alpha_0}) = X + Y$. Geometrically, $X + Y$ is the set of all points which are in X or in Y but not in both.

For example, if $X = (1, 0, 0, 1)$ and $Y = (0, 1, 0, 1)$, $X + Y = (1, 1, 0, 0)$; i.e., X represents a_1^0 and a_4^0, Y represents a_2^0 and a_4^0, and $X + Y$ represents a_1^0 and a_2^0. Since a_4^0 appears in both X and Y, it is suppressed in forming the sum, modulo 2.

This type of addition has the obvious property that if two sets contain each an even number of 0-cells, the sum (mod. 2) contains an even number of 0-cells.

15. Any set, S, of 1-cells in C_1 may be denoted by a symbol $(x_1, x_2, \ldots, x_{\alpha_1})$ in which $x_i = 1$ if α_i^1 is in the set and $x_i = 0$ if α_i^1 is not in the set. The 1-cells in the set may be thought of as labelled with 1's and those not in the set as labelled with 0's. The symbol is also regarded as representing the one-dimensional complex composed of the 1-cells of S and the 0-cells which bound them. Thus, for example, in fig. [8.3] the boundaries of two of the faces are $(1, 0, 1, 0, 1, 0)$ and $(1, 1, 0, 0, 0, 1)$.

The sum (mod. 2) of two symbols $(x_1, x_2, \ldots, x_{\alpha_1})$ is defined in the same way as for the case of symbols representing 0-cells. Correspondingly if C_1' and C_1'' are one-dimensional complexes which have a certain number (which may be zero) of 1-cells in common and have no other common points except the ends of these 1-cells, the *sum*

$$C_1' + C_1'' \text{ (mod. 2)}$$

is defined as the one-dimensional complex obtained by suppressing all 1-cells common to C_1' and C_1'' and retaining all 1-cells which appear only in C_1' or in C_1''. For example, in Fig. [8.3], the sum of the two curves represented by $(1, 0, 1, 0, 1, 0)$ and $(1, 1, 0, 0, 0, 1)$ is $(0, 1, 1, 0, 1, 1)$ which represents the curve composed of $a_2{}^1$, $a_3{}^1$, $a_5{}^1$, $a_6{}^1$ and their ends.

The Matrices H_0 and H_1

16. ..... Any one-dimensional complex falls into R_0 sub-complexes each of which is connected. Let us denote these sub-complexes by $C_1{}^1, C_1{}^2, \ldots, C_1{}^{R_0}$, and let the notation be assigned in such a way that $a_i{}^0$ $(i = 1, 2, \ldots, m_1)$ are the 0-cells of $C_1{}^1$, $a_i{}^0$ $(i = m_1 + 1, \ldots, m_2)$ those of $C_1{}^2$, and so on.

With this choice of notation, the sets of vertices of $C_1{}^1, C_1{}^2, \ldots, C_1{}^{R_0}$, respectively, are represented by the symbols $(x_1, x_2, \ldots, x_{\alpha_0})$ which constitute the rows of the following matrix.

$$
H_0 = \left\| \begin{array}{ccccccccccccc}
\overbrace{1 & 1 & \cdots & 1}^{m_1} & \overbrace{0 & 0 & \cdots & 0}^{m_2 - m_1} & \cdots & \overbrace{0 & 0 & \cdots & 0}^{\alpha_0 - m_{R_0 - 1}} \\
0 & 0 & \cdots & 0 & 1 & 1 & \cdots & 1 & \cdots & 0 & 0 & \cdots & 0 \\
& & & & & & & & & & & & \\
& & & & & & & & & & & & \\
0 & 0 & \cdots & 0 & 0 & 0 & \cdots & 0 & \cdots & 1 & 1 & \cdots & 1
\end{array} \right\| = \| \eta_{ij}{}^0 \|
$$

For most purposes it is sufficient to limit attention to connected complexes. In such cases $R_0 = 1$, and H_0 consists of one row all of whose elements are 1.

17. By the definition ..... a 0-cell is *incident* with a 1-cell if it is one of the ends of the 1-cell, and under the same conditions the 1-cell is incident with the 0-cell. The incidence relations between the 0-cells and the 1-cells may be represented in a table or matrix of α_0 rows and α_1 columns as follows: The 0-cells of C_1 having been denoted by $a_i{}^0$, $(i = 1, 2, \ldots, \alpha_0)$ and the 1-cells by $a_j{}^1$, $(j = 1, 2, \ldots, \alpha_1)$, let the element of the ith row and the jth column of the matrix be 1 if $a_i{}^0$ is incident with $a_j{}^1$ and let it be 0 if $a_i{}^0$ is not incident with $a_j{}^1$.

For example, the table for the linear graph of Fig. [8.3] formed by the vertices and edges of a tetrahedron is as follows:

	$a_1{}^1$	$a_2{}^1$	$a_3{}^1$	$a_4{}^1$	$a_5{}^1$	$a_6{}^1$
$a_1{}^0$	1	0	0	0	1	1
$a_2{}^0$	0	1	0	1	0	1
$a_3{}^0$	0	0	1	1	1	0
$a_4{}^0$	1	1	1	0	0	0

In the case of the complex used ..... to define a simple closed curve the incidence matrix is

$$
\left\| \begin{array}{cc} 1 & 1 \\ 1 & 1 \end{array} \right\| .
$$

We shall denote the element of the ith row and jth column of the matrix of incidence relations between the 0-cells and 1-cells by $\eta_{ij}{}^1$ and the matrix itself by

$$
\| \eta_{ij}{}^1 \| = H_1.
$$

The ith row of H_1 is the symbol for the set of all 1-cells incident with a_i^0 and the jth column is the symbol for the set of two 0-cells incident with a_j^1.

The condition which we have imposed on the graph, that both ends of every 1-cell shall be among the α_0 0-cells, implies that every column of the matrix contains exactly two 1's. Conversely, any matrix whose elements are 0's and 1's and which is such that each column contains exactly two 1's can be regarded as the incidence matrix of a linear graph. For to obtain such a graph it is only necessary to take α_0 points in a 3-space, denote them arbitrarily by $a_1^0, a_2^0, \ldots,$ $a_{\alpha_0}^0$, and join the pairs which correspond to 1's in the same column successively by arcs not meeting the arcs previously constructed.

*   *   *   *   *

20. Denoting the connected sub-complexes of C_1 by $C_1^1, C_1^2, \ldots, C_1^{R_0}$ as in §16 let the notation be so assigned that $a_1^1, \ldots, a_{m_1}^1$ are the 1-cells in C_1^1, $a_{m_1+1}^1, \ldots,$ $a_{m_2}^1$ the 1-cells in C_1^2; and so on. The matrix H_1 then must take the form

$$\left\| \begin{array}{c:c:c:c} I & 0 & 0 & 0 \\ \hdashline 0 & II & 0 & 0 \\ \hdashline 0 & 0 & III & \\ \hdashline & & & \end{array} \right\|$$

where all the non-zero elements are to be found in the matrices I, II, III, etc., and I is the matrix of C_1^1, II of C_1^2, etc. This is evident because no element of one of the complexes C_1^i is incident with any element of any of the others.

There are two non-zero elements in each column of H_1. Hence if we add the rows corresponding to any of the blocks I, II, etc. the sum is zero (mod. 2) in every column. Hence the rows of H_1 are connected by R_0 linear relations.

Any linear combination (mod. 2) of the rows of H_1 corresponds to adding a certain number of them together. If this gave zeros in all the columns it would mean that there were two or no 1's in each column of the matrix formed by the given rows, and this would mean that any 1-cell incident with one of the 0-cells corresponding to these rows would also be incident with another such 0-cell. These 0-cells and the 1-cells incident with them would therefore form a sub-complex of C_1 which was not connected with any of the remaining 0-cells and 1-cells of C_1. Hence it would consist of one or more of the complexes C_1^i $(i = 1, 2, \ldots, R_0)$ and the linear relations with which we started would be dependent on the R_0 relations already found. Hence there are exactly R_0 linearly independent linear relations among the rows of H_1, so that if ρ_1 is the rank of H_1,

$$\rho_1 = \alpha_0 - R_0.$$

*   *   *   *   *

One-dimensional Circuits

22. A connected linear graph each vertex of which is an end of two and only two 1-cells is called a *one-dimensional circuit* or a *1-circuit*. ..... Any closed curve is decomposed by any finite set of points on it into a 1-circuit. Conversely, it is easy to see that the set of all points on a 1-circuit is a simple closed curve. It is obvious, further, that any linear graph such that each vertex is an end of two and only two 1-cells is either a 1-circuit or a set of 1-circuits no two of which have a point in common.

Consider a linear graph C_1 such that each vertex is an end of an even number of edges. Let us denote by $2n_i$ the number of edges incident with each vertex a_i^0. The edges incident with each vertex a_i^0 may be grouped arbitrarily in n_i pairs no two of which have an edge in common; let these pairs of edges be called the pairs associated with the vertex a_i^0. Let C_1' be a graph coincident with C_1 in such a way that (1) there is one and only one point of C_1' on each point of C_1 which is not a vertex and (2) there are n_i vertices of C_1' on each vertex a_i^0 of C_1 each of these vertices of C_1' being incident only with the two edges of C_1' which coincide with a pair associated with a_i^0.

The linear graph C_1' has just two edges incident with each of its vertices and therefore consists of a number of 1-circuits. Each of these 1-circuits is coincident with a 1-circuit of C_1, and no two of the 1-circuits of C_1 thus determined have a 1-cell in common. Hence C_1 consists of a number of 1-circuits which have only a finite number of 0-cells in common.

It is obvious that a linear graph composed of a number of closed curves having only a finite number of points in common has an even number of 1-cells incident with each vertex. Hence *a necessary and sufficient condition that C_1 consist of a number of 1-circuits having only 0-cells in common is that each 0-cell of C_1 be incident with an even number of 1-cells.* A set of 1-circuits having only 0-cells in common will be referred to briefly as a set of 1-circuits.

*   *   *   *   *

24. Let us now inquire under what circumstances a symbol $(x_1, x_2, \ldots, x_{\alpha_1})$ for a one-dimensional complex contained in C_1 will represent a 1-circuit or a system of 1-circuits.

Consider the sum

$$\eta_{i1}{}^1 x_1 + \eta_{i2}{}^1 x_2 + \ldots + \eta_{i\alpha_1}{}^1 x_{\alpha_1}$$

where the coefficients $\eta_{ij}{}^1$ are the elements of the ith row of H_1. Each term $\eta_{ij}{}^1 x_j$ of this sum is 0 if a_j^1 is not in the set of 1-cells represented by $(x_1, x_2, \ldots, x_{\alpha_1})$ because in this case $x_j = 0$; it is also zero if a_j^1 is not incident with a_i^0 because $\eta_{ij}{}^1 = 0$ in this case. The term $\eta_{ij}{}^1 x_j = 1$ if a_j^1 is incident with a_i^0 and in the set represented by $(x_1, x_2, \ldots, x_{\alpha_1})$ because in this case $\eta_{ij}{}^1 = 1$ and $x_j = 1$. Hence there are as many non-zero terms in the sum as there are 1-cells represented by $(x_1, x_2, \ldots, x_{\alpha_1})$ which are incident with a_i^0. Hence by §22 the required condition is that the number of non-zero terms in the sum must be even. In other words if the x's and $\eta_{ij}{}^1$'s are reduced modulo 2 as explained in §14 we must have

$$(\mathrm{H}_1) \qquad\qquad \sum_{j=1}^{\alpha_1} \eta_{ij}{}^1 x_j = 0 \qquad\qquad (i = 1, 2, \ldots, \alpha_0)$$

if and only if $(x_1, x_2, \ldots, x_{\alpha_1})$ represents a 1-circuit or set of 1-circuits. The matrix of this set of equations (or congruences, mod. 2) is H_1.

25. If the rank of the matrix H_1 of the equations (H_1) be ρ_1 the theory of linear homogeneous equations (congruences, mod. 2) tells us that there is a set of $\alpha_1 - \rho_1$ linearly independent solutions of (H_1) upon which all other solutions are linearly dependent. This means geometrically that *there exists a set of $\alpha_1 - \rho_1$ 1-circuits or systems of 1-circuits from which all others can be obtained by repeated applications of the operation of adding (mod. 2) described in §14.* We shall call this a *complete set* of 1-circuits or systems of 1-circuits.

Since $\rho_1 = \alpha_0 - R_0$ (§ 20), the number of solutions of (H_1) in a complete set is

$$\mu = \alpha_1 - \alpha_0 + R_0,$$

where μ is the cyclomatic number ..... For the sake of uniformity with a notation used later on we shall also denote μ by $R_1 - 1$. Thus we have

$$\alpha_0 - \alpha_1 = 1 + R_0 - R_1.$$

Trees

26. A connected linear graph which contains no 1-circuits is called a *tree*. As a corollary of the last section it follows that *a linear graph is a set of R_0 trees if and only if $\mu = 0$.*

Any connected linear graph C_1 can be reduced to a tree by removing μ properly chosen 1-cells. For let a_p^1 $(p = i_1, i_2, \ldots, i_{p_1})$ be a set of 1-cells whose boundaries form a complete set of 0-circuits (§ 20). The remaining 1-cells of C_1 are μ in number and will be denoted by a_p^1 $(p = j_1, j_2, \ldots, j_\mu)$. If these μ 1-cells are removed from C_1 the linear graph T_1 which remains is connected because every bounding 0-circuit of C_1 is linearly expressible in terms of the boundaries of the 1-cells a_p^1 $(p = i_1, i_2, \ldots, i_{p_1})$ of T_1 and hence any two 0-cells of C_1 are joined by a curve composed of 1-cells of T_1. But since the cyclomatic number of C_1 is $\mu = \alpha_1 - \alpha_0 + 1$, the removal of μ 1-cells reduces it to 0 and hence reduces C_1 to a tree. In like manner, if C_1 is a linear graph for which $R_0 > 1$, it can be reduced to R_0 trees by removing $\mu = \alpha_1 - \alpha_0 + R_0$ properly chosen 1-cells.

27. There is at least one 1-circuit of C_1 which contains the 1-cell $a_{j_1}^1$, for otherwise C_1 would be separated into two complexes by removing this 1-cell. Call such a 1-circuit C_1^1. In the complex obtained by removing $a_{j_1}^1$ from C_1 there is, for the same reason, a 1-circuit C_1^2 which contains $a_{j_2}^1$, and so on. Thus there is a set of 1-circuits $C_1^1, C_1^2, \ldots, C_1^\mu$ such that C_1^p $(p = 1, 2, \ldots, \mu)$ contains $a_{j_p}^1$. These 1-circuits are linearly independent because C_1^{k-1} contains a 1-cell, $a_{j_k}^1$, which does not appear in any of the circuits $C_1^\mu, C_1^{\mu-1}, \ldots, C_1^k$ and therefore cannot be linearly dependent on them. Hence $C_1^1, C_1^2, \ldots, C_1^\mu$ constitute a complete set of 1-circuits. This sharpens the theorem of § 25 a little in that it establishes that there is a complete set of solutions of (H_1) each of which represents a single 1-circuit.

Planar graphs

Although our next extract is also concerned with a topological problem, the treatment therein is not based on algebraic methods, but belongs rather to the realm of 'analytic' topology. This subject flourished in the first decades of this century, one of its highlights being the proof of what is now known as the Jordan curve theorem. This theorem states that a plane curve which is closed and does not intersect itself divides the plane into two connected sets, a bounded 'inside' and an unbounded 'outside'. Jordan was the first to recognize that such an intuitively obvious result required proof, but his own attempt was inadequate, and the first correct proof was given by Veblen in 1905 [4]. From our point of view, the interest lies in the application of this kind of result to planar graphs.

We recall from Chapter 5 that a graph is planar if its vertices and edges can be represented by points and lines in the plane, in such a way that the lines do not cross. Like several other aspects of graph theory, the study of planarity originated from puzzles. One such puzzle is the problem of Möbius which we discussed in the previous chapter, and which we shall reconsider here in a dual form.

Let us suppose that the capital cities of the five neighbouring regions are to be joined in pairs by roads, in such a way that no bridges or crossroads are necessary. The five cities and ten roads may then be regarded as the vertices and edges of the complete graph K_5, and the problem requires us to draw this graph in the plane without crossings. A few experiments with pencil and paper will convince the reader that the problem is insoluble, and that K_5 is consequently not planar. We shall give a proof of this result shortly.

Another puzzle involving planarity was included in a book of *Amusements in mathematics* [5], compiled by H. E. Dudeney and first published in 1917. Dudeney presented the problem, under the title 'Water, gas, and electricity', in the following way.

> There are some half-dozen puzzles, as old as the hills, that are perpetually cropping up, and there is hardly a month in the year that does not bring inquiries as to their solution. Occasionally one of these, that one had thought was an extinct volcano, bursts into eruption in a surprising manner. I have received an extraordinary number of letters respecting the ancient puzzle that I have called 'Water, Gas, and Electricity'. It is much older than electric lighting, or even gas, but the new dress brings it up to date. The

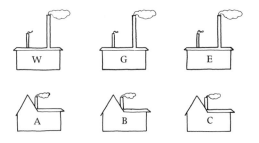

> puzzle is to lay on water, gas, and electricity, from W, G, and E, to each of the three houses, A, B, and C, without any pipe crossing another. Take your pencil and draw lines showing how this should be done. You will soon find yourself landed in difficulties.

Although Dudeney spoke of the puzzle as an ancient one, we have been unable to find any earlier printed references to it. Sam Loyd Jnr. claimed [6] that his father 'brought out' the puzzle in 1900, but he did not say who had invented it.

Graph theorists now have a special name for the kind of graph encountered in this puzzle. The **complete bipartite graph** $K_{s,t}$ has two sets of vertices, one with s members and the other with t members; each

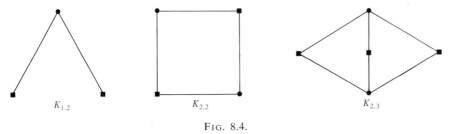

$$K_{1,2} \qquad K_{2,2} \qquad K_{2,3}$$

FIG. 8.4.

vertex in the first set is joined to each vertex in the second set, so that there are st edges in all. For example, the graphs $K_{1,2}$, $K_{2,2}$, and $K_{2,3}$ are depicted in Fig. 8.4, with circles and squares indicating the two sets of vertices; these graphs are immediately seen to be planar, since the plane drawings of them have no crossings. But the graph $K_{3,3}$, which is the subject of the gas, water, and electricity puzzle, is non-planar, as we shall now prove.

The proofs for both K_5 and $K_{3,3}$ use Euler's formula. (For reference, diagrams of these graphs, with some unavoidable crossings, are given in Fig. 8.5.) Let us first consider K_5. In any faithful planar representation of it there would have to be N_f regions, where N_f satisfies Euler's formula $N_v - N_e + N_f = 2$. In this case $N_v = 5$ and $N_e = 10$, so $N_f = 7$. Now each region must be bounded by at least three edges (since there are no loops or multiple edges), and each edge belongs to the boundary of just two regions. Consequently, for seven regions we need at least $\frac{1}{2}(7 \times 3)$ edges; since this number exceeds 10, K_5 cannot be planar. A similar argument applies to $K_{3,3}$; here $N_v = 6$ and $N_e = 9$, so that $N_f = 5$. But each circuit in $K_{3,3}$ has at least four edges, and so the boundary of each region must contain at least four edges. Thus we need at least $\frac{1}{2}(5 \times 4)$ edges altogether, and since this number exceeds 9, $K_{3,3}$ is not planar.

The reader may be wondering why so much discussion has been devoted to these two particular examples of non-planar graphs. The reason is that in 1930 the Polish mathematician Kazimierz KURATOWSKI published a remarkable theorem, in the form of a converse to the simple

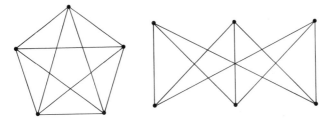

FIG. 8.5.

results just described. He proved that if a graph is non-planar, then it must contain either K_5 or $K_{3,3}$; in other words, these two graphs are essentially the only obstacles to planarity. According to Kuratowski himself [7], when he began to think about the problem of non-planar graphs he expected that K_5 would be the only obstacle. Clearly, he was not familiar with the gas, water, and electricity puzzle!

In our statement of Kuratowski's theorem, we did not explain just what is meant by the word 'contain'. To do so, we use the idea of subdividing an edge of a graph—that is, replacing an edge joining the vertices v and w by a chain of edges, from v to v_1, v_1 to $v_2, \ldots, v_{r-1}$ to v_r, and v_r to w, where each of the new vertices v_i has valency two. If G is a given graph and H is a graph derived from it by subdividing some of the edges, then H is said to be a **subdivision** of G. Any two subdivisions of G are equivalent from a topological point of view, and they are said to be **homeomorphic** to each other, and to G. The precise statement of Kuratowski's theorem says that *every non-planar graph has a subgraph homeomorphic to either K_5 or $K_{3,3}$*. Typical subdivisions of the basic graphs are depicted in Fig. 8.6.

The next extract [8C] comes from the paper in which Kuratowski proved his theorem, and it contains the essential part of the proof. The original (in French) was written from the standpoint of analytic topology, and so it is necessary to begin by explaining several points concerning terminology and notation. Kuratowski dealt with a structure rather more general than a graph, called a *continu Péanien*; we lose nothing by translating this phrase by the word 'graph', provided that we remember that it signifies a topological realization of a graph as a system of points and lines. The term *gauche* is accurately translated as 'non-planar'.

The first few sections of the paper were devoted to the proof of some intuitively obvious results whose significance is topological, rather than graph-theoretical. We ask the reader to take for granted the meaning of such terms as 'inside', 'outside', and 'boundary'; if these words are given their usual interpretation, then the paper becomes more easily intelligible, without losing the essence of the argument.

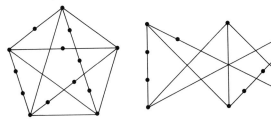

FIG. 8.6.

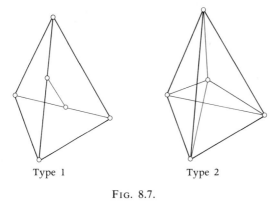

Type 1 Type 2

FIG. 8.7.

The notation used in [8C] needs a little explanation. A path joining two points a and b, without its end points, is denoted by ab; when it lies in a specified set S, it is denoted by $(ab)_S$. A circuit in S, through a and b, gives rise to two paths joining a and b; they are distinguished by giving the circuit an orientation, and writing them as $(ab)_S$ and $(ba)_S$ to conform with the orientation. The symbol $\bar{X}$ is used to denote a set X together with its boundary points.

After setting up the necessary topological machinery, Kuratowski went on to state his main theorem:

> Theorem A. If A is a connected non-planar graph which has only a finite number of circuits, then A contains a subgraph homeomorphic to the graph of type 1 or the graph of type 2. [Fig. 8.7.]

The graph of type 1 is $K_{3,3}$, and the graph of type 2 is K_5. In order to simplify the proof of Theorem A, Kuratowski defined two vertices x and y of a graph C to be *conjugate* if there is a planar representation of C in which x and y can be joined by a line which does not meet the rest of C. He then considered the graph M formed from the non-planar graph A by removing an edge joining two vertices, a and b. He showed that if (i) there is no circuit in M through a and b, or (ii) a and b are conjugate

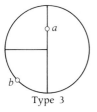

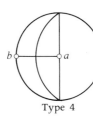

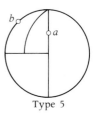

Type 3   Type 4   Type 5

FIG. 8.8.

relative to M, then the hypothesis that A is non-planar is contradicted. So the critical case is when M does not have these two properties.

Kuratowski remarked that it is sufficient to show that, in this case, M must contain a subgraph Q of one of the three types depicted in Fig. 8.8. For if M contains Q, then $A = M \cup ab$ contains $Q \cup ab$: if Q is of type 3 then $Q \cup ab$ is of type 1; if Q is of type 4 then $Q \cup ab$ is of type 2; and if Q is of type 5 then $Q \cup ab$ (less the path joining b to the vertex of valency 4) is of type 1 again. Thus the proof of Theorem A is reduced to the proof of what Kuratowski called Theorem B.

8C

K. KURATOWSKI
SUR LE PROBLÈME DES COURBES GAUCHES EN TOPOLOGIE
[On the topological problem of non-planar curves]

Fundamenta Mathematicae **15** (1930), 271–283.

*   *   *   *   *

Theorem B. *Given a planar representation of a graph M containing only a finite number of circuits and such that* 1: *M contains two non-conjugate vertices a and b,* 2: *a and b lie on a circuit in M,—then M contains a graph homeomorphic to one of type* 3, 4 *or* 5.

5. Proof of theorem B.

Among the circuits contained in M and containing a and b, there is one circuit K with the property that there exists no other circuit K' whose inside is contained in V, the inside of K.

For, otherwise, there would be an infinity of circuits in M.

K being so defined, *if cd is a path lying in* $V \cap M$ *and having its extremities on K, then one of them belongs to* $(ab)_K$ *and the other to* $(ba)_K$ (these symbols denoting the paths in K).

Let W be outside K.

I shall prove that, among the components of $M \cap W$, there is one component S such that:

1: $\bar{S}$ contains two vertices p and q such that $p \in (ab)_K$ and $q \in (ba)_K$,

2: $\bar{S} \cap K$ contains two points s and t which cannot be joined by any path $st \subset V - M$.

To begin, I shall prove the existence of a component S satisfying condition 1 and the following:

3: there is no component R of $V - M$ such that $\bar{S} - S \subset \bar{R}$.

Let us suppose, in order to obtain a contradiction, that there is no component S satisfying conditions 1 and 3. Let G be the union of all the S such that $\bar{S} \cap (ab)_K \neq \varnothing \neq \bar{S} \cap (ba)_K$. The family of all these S is infinite, ... and consequently, there exists a finite sequence $R_1, \ldots, R_n$ of components of $V - M$ such that, for each S, the set $\bar{S} - S$ is contained in one of the sets $\bar{R}_i$. Let $G_1 =$ the union of all the S such that $\bar{S} - S \subset \bar{R}_1$, and, in general, let $G_i =$ the union of the S such that $\bar{S} - S \subset \bar{R}_i$ and S is not in any G_j with $j < i$. Thus $G = G_1 + \ldots + G_n$, $\bar{G}_i - G_i \subset \bar{R}_i$ and the G_i are disjoint. $\ldots\ldots$ It follows that M is homeomorphic to a graph $M_1 = G_1^* \cup M - G_1$, where $G_1^* \subset R_1$.

Now the same proposition can also be applied to the graph M_1. In fact, the set G_2, since it is open in M, is open in M_1, since $G_2 \subset W$ and $G_1^* \subset V$. On the other hand, R_2, being a component of $V - M$, is a component of $V - M_1$, since the inclusion $G_1^* \subset R_1$ implies $\bar{R}_2 \cap M = \bar{R}_2 \cap M_1$.

Thus we may substitute M_1 for M, R_2 for R, and G_2 for G into the preceding proposition. Continuing in the same way we arrive at the conclusion that M is homeomorphic to a graph $M^* = G^* \cup M - G$ where $G^* \subset V$.

Since $M^* \cap W = M \cap W - G$, we conclude from the definition of G that $M^* \cap W$ contains no connected set (and, more particularly, no component) S' such that $\bar{S}' \cap (ab)_K \neq \varnothing \neq \bar{S}' \cap (ba)_K$. But then $\ldots\ldots$ the points a and b may be joined by a path ab disjoint from M^*, contrary to the hypothesis that these two points are non-conjugate relative to M.

The existence of a component S which satisfies conditions 1 and 3 is thereby established.

This component also satisfies condition 2. This is evident in the case where the set $\bar{S} - S$ reduces to two points. So let us suppose that $\bar{S} - S$ contains three or more points, of which two, x and y, belong to $(\overline{ab})_K$ and, furthermore, suppose that S satisfies condition 3 but not 2. $\ldots\ldots$ The existence of a path $xy \subset V \cap M$ follows. Now the ends of this path are situated on $(\overline{ab})_K$, leading to a contradiction of the property of K established at the start of the proof.

Thus it is established that S satisfies 2.

Condition 2 implies $\ldots\ldots$ the existence of a path $cd \subset V \cap M$ such that $c \in (st)_K$ and $d \in (ts)_K$. Furthermore, according to the property of K just mentioned, we may assume that $c \in (ab)_K$ and $d \in (ba)_K$.

We shall now pick out from $K \cup S \cup cd$ a graph which is of type 3, 4, or 5. We shall have 4 cases to distinguish:

1) $s \in (ab)_K$ and $t \in (ba)_K$.

In this case, the graph $K \cup (st)_S \cup cd$ is of type 3.

Thus it is permissible to assume in the following cases that there exists no pair s, t which satisfies conditions 2 and 1) at the same time.

2) $s = a$, $t \in (ba)_K$.

Let T be a "T-shaped" graph (triod) taken from S and having the vertices p, s, t for its ends. The graph $K - (ac)_K \cup T \cup cd$ is of type 3.

3) $s = a$, $t = b$. Thus $p = c$ and $q = d$ (because otherwise we are in case 1) or 2)).

If there exists in S an "X-shaped" graph having for its ends the points a, b, c, d, then the graph $K \cup X \cup cd$ is of type 4. If this is not so, let T be the triod defined as above. Let v be its "centre". In $S - T$ there is a path D which joins d to T; this path leads either to av, or to bv. Evidently we may assume that the the former is the case; the graph $K - (cb)_K - (da)_K \cup T \cup D \cup cd$ is of type 3.

4) The only case which remains to be considered is: $s, t \in (ba)_K$ and $p = c$.

In this case, the graph $K \cup T \cup cd$ is of type 5.

Theorem **B** and, consequently, theorem **A** are thereby proved.

* * * * *

Kuratowski announced his results to the Polish Mathematical Society in Warsaw on 21 June 1929. A few months later, before his paper had appeared in print, the following abstract was published in the *Bulletin of the American Mathematical Society* [8].

> *Irreducible non-planar graphs.* One of the results of this paper is a simple necessary and sufficient condition that an arbitrary linear graph be mappable on a plane. (Received February 10, 1930.)

In fact, the authors of this paper, Orrin FRINK and P. A. Smith, had independently arrived at the same result as Kuratowski. Their paper was intended for publication in the *Transactions of the American Mathematical Society*, but, as Frink informed us in a letter written in 1974: 'Unfortunately Kuratowski's proof came out in *Fundamenta* just at that time, and equally unfortunate was the fact that our proof was similar to Kuratowski's. Hence our paper was simply rejected by the *Transactions*.'

Such are the pitfalls of original research. However, Frink and Smith could at least console themselves with the thought that the credit had gone to the first person to announce the result. History has not been so kind to Kirkman and his work on 'Hamiltonian' circuits.

Planarity and Whitney duality

The notion that planar graphs could be characterized by a property which is combinatorial, rather than topological, had been suggested long before Kuratowski obtained his theorem. In 1916 Dénes König [9] declared that progress on the four-colour conjecture might depend upon such a characterization. Indeed, although the famous conjecture played no part in Kuratowski's discovery, it was indirectly responsible for a similar result shortly afterwards. This result was due to a young American, Hassler WHITNEY, who had become interested in the four-colour problem in the late 1920s, while studying physics in Germany. In the years from 1930 to 1935 he wrote about a dozen papers on various aspects of graph theory. Some of these were concerned with a combinatorial characterization of planarity, and we shall give an account of them here; others, more directly relevant to colouring problems, will be surveyed in the next chapter.

Whitney's approach to planarity was based on the notion of duality. In Chapter 7 we explained how a map M on a surface gives rise to a dual map M^* on the same surface, by means of a simple geometrical construction. In the special case of a map formed by a graph G drawn in the plane, the vertices and edges of the dual map form another planar graph G^*, which we shall call the **geometric dual** of G. It must be stressed that the idea of duality is now being applied to graphs on a particular surface (the plane), rather than to maps. An example is illustrated in Fig. 8.9, where the edges of G are denoted by continuous lines and those of G^* by broken lines. There is a one-to-one correspondence between the edges of

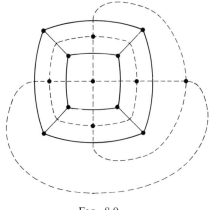

FIG. 8.9.

G and the edges of G^*, which we emphasize in the illustration by ensuring that corresponding edges cross one another. If G is connected, then $G^{**} = G$; but this is not so when G has more than one component, as shown by the example in Fig. 8.10.

The geometric duality between G and G^* leads to some combinatorial relationships between them, when they are considered as abstract graphs. Whitney studied these relationships and used them to formulate an abstract notion of duality; if G and G^* are geometric duals, then they are duals in Whitney's sense also. Furthermore, it turns out that this abstract notion is precisely equivalent to the geometric one in the planar case: a graph is planar if and only if it has a Whitney dual. This result, which first appeared in 1931 [10], is Whitney's combinatorial characterization of planarity.

In 1932 a full account of Whitney's work on planarity was published, and our next extract [8D] is taken from this paper. The first part of the paper dealt with the process of separating a graph into what Whitney called its 'components'. Unfortunately, he used this term in a non-standard way. We say that a graph is **non-separable** if it is connected and cannot be disconnected by removing a single vertex (a **cut vertex**). In Whitney's sense, a 'component' is a maximal non-separable part of a graph. Nowadays we call this a **block.** For example, the graph on the left

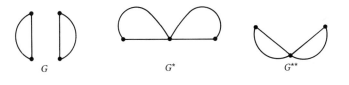

FIG. 8.10.

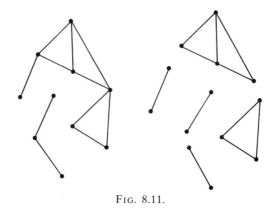

FIG. 8.11.

of Fig. 8.11 has five such 'components' (blocks), as shown in the diagram on the right. A graph with more than one block is said to be **separable**. Whitney used the term 'connected piece' to denote a component in the usual sense of the word; the graph on the left of Fig. 8.11 has just two 'connected pieces'.

The theorems about separability in graphs are of a routine kind, and in our extract [8D] we have omitted a lengthy discussion of them. Those that are needed for the proof of the main result are stated here for reference, with Whitney's numbering and terminology.

> THEOREM 6. Let G be a connected graph containing no 1-circuit [loop]. A necessary and sufficient condition that the vertex a be a cut vertex of G is that there exist two vertices b, c in G, each distinct from a, such that every chain from b to c passes through a.

> THEOREM 8. A non-separable graph G containing at least two arcs [edges] contains no 1-circuit and is of nullity >0. Each vertex is on at least two arcs.

> THEOREM 9. Let G be a graph of nullity 1 containing no isolated vertices, such that the removal of any arc reduces the nullity to 0. Then G is a circuit.

> THEOREM 18. If G is a non-separable graph of nullity $N>1$, we can remove an arc or suspended chain [subdivided edge] from G, leaving a non-separable graph G' of nullity $N-1$.

A result concerning the relationship between separability and duality is also required:

> THEOREM 25. Let G and G' be dual graphs, and let $H_1, \ldots, H_m$ be the components of G. Let $H'_1, \ldots, H'_m$ be the corresponding subgraphs of G'. Then $H'_1, \ldots, H'_m$ are the components of G', and H'_i is a dual of H_i, $i=1, \ldots, m$.

In particular,

> THEOREM 26. A dual of a non-separable graph is non-separable.

8D

H. WHITNEY
NON-SEPARABLE AND PLANAR GRAPHS

Transactions of the American Mathematical Society **34** (1932), 339–362.

$$* \quad * \quad * \quad * \quad *$$

2. **Rank and nullity.** Given a graph G which contains V vertices, E arcs, and P connected pieces, we define its *rank R*, and its *nullity* (or *cyclomatic number* or first Betti number) N, by the equations

$$R = V - P,$$
$$N = E - R = E - V + P.$$

(These are just the rank and nullity of the matrix H_1 of Poincaré. See Veblen's Colloquium Lectures, *Analysis Situs* [8B].)

If G contains the single arc ab, it is of rank 1, nullity 0, while if it contains the single arc aa, it is of rank 0, nullity 1.

The first two theorems follow immediately from the definitions of rank and nullity:

THEOREM 1. *If isolated vertices be added to or subtracted from a graph, the rank and nullity remain unchanged.*

THEOREM 2. *Let the graph G' be formed from the graph G by adding the arc ab. Then*

(1) *if a and b are in the same connected piece in G, then*

$$R' = R, N' = N + 1;$$

(2) *if a and b are in different connected pieces in G, then*

$$R' = R + 1, N' = N.$$

THEOREM 3. *In any graph G,*

$$R \geqq 0, N \geqq 0.$$

For let G_1 be the graph containing the vertices of G but no arcs. Then if R_1 and N_1 are its rank and nullity,

$$R_1 = N_1 = 0.$$

We build up G from G_1 by adding the arcs one at a time. The theorem now follows from Theorem 2.

$$* \quad * \quad * \quad * \quad *$$

8. **Duals.** Given a graph G, if H_1 is a subgraph of G, and H_2 is that subgraph of G containing those arcs not in H_1, we say H_2 is the *complement* of H_1 in G.

Throughout this section, R, R', r, r', etc., will stand for the ranks of G, G', H, H', etc., respectively, with similar definitions for V, E, P, N.

Definition. Suppose there is a (1, 1) correspondence between the arcs of the graphs G and G', such that if H is any subgraph of G and H' is the complement of the corresponding subgraph of G', then

$$r' = R' - n.$$

We say then that G' is a *dual* of G.

(While this definition agrees with the ordinary one for graphs lying on a plane or sphere, a graph on a surface of higher connectivity, such as the torus, has in general no dual. See Theorems 29 and 30.)

Thus if the nullity of H is n, then H' (including all the vertices of G') is in n more connected pieces than G'.

THEOREM 20. *Let G' be a dual of G. Then*

$$R' = N,$$

$$N' = R.$$

For let H be that subgraph of G consisting of G itself. Then

$$n = N.$$

If H' is the complement of the corresponding subgraph of G', H' contains no arcs, and is the null graph. Thus

$$r' = 0.$$

But as G' is a dual of G,

$$r' = R' - n.$$

These equations give

$$R' = N.$$

The other equation follows when we note that $E' = E$.

THEOREM 21. *If G' is a dual of G, then G is a dual of G'.*

Let H' be any subgraph of G', and let H be the complement of the corresponding subgraph of G. Then, as G' is a dual of G,

$$r' = R' - n.$$

By Theorem 20, $$R' = N.$$

We note also,

$$e + e' = E.$$

These equations give

$$r = e - n = e - (R' - r') = e - N + (e' - n')$$
$$= E - N - n' = R - n'.$$

Thus G is a dual of G'.

Whenever we have shown that one graph is a dual of another graph, we may now call the graphs "dual graphs".

*   *   *   *   *

9. **Planar graphs.** Up till now, we have been considering abstract graphs alone. However, the definition of a planar graph is topological in character. This section may be considered as an application of the theory of abstract graphs to the theory of topological graphs.

Definitions. A topological graph is called *planar* if it can be mapped in a (1, 1) continuous manner on a sphere (or a plane). For the present, we shall say that an abstract graph is *planar* if the corresponding topological graph is planar. Having proved Theorem 29, we shall be justified in using the following purely combinatorial definition: *A graph is planar if it has a dual.*

We shall henceforth talk about "graphs" simply, the terms applying equally well to either abstract or topological graphs.

LEMMA. *If a graph can be mapped on a sphere, it can be mapped on a plane, and conversely.*

Suppose we have a graph mapped on a sphere. We let the sphere lie on the plane, and rotate it so that the new north pole is not a point of the graph. By stereographic projection from this pole, the graph is mapped on the plane. The inverse of this projection maps any graph on the plane onto the sphere.

By the *regions* of a graph lying on a sphere or in a plane is meant the regions into which the sphere or plane is thereby divided. A given region of the graph is characterized by those arcs of the graph which form its boundary. If the graph is in a plane, the outside region is the unbounded region.

LEMMA. *A planar graph may be mapped on a plane so that any desired region is the outside region.*

We map the graph on a sphere, and rotate it so that the north pole lies inside the given region. By stereographic projection, the graph is mapped onto the plane so that the given region is the outside region.

We return now to the work in hand.

THEOREM 27. *If the components of a graph G are planar, G is planar.*

Suppose the graphs G_1 and G_2 are planar, and G' is formed by letting the vertices a_1 and a_2 of G_1 and G_2 coalesce. We shall show that G' is planar. Map G_1 on a sphere, and map G_2 on a plane so that one of the regions adjacent to the vertex a_2 is the outside region. Shrink the portion of the plane containing G_2 so it will fit into one of the regions of G_1 adjacent to a_1. Drawing a_1 and a_2 together, we have mapped G' on the sphere. (Here and in a few other places we are using point-set theorems which, however, are geometrically evident.) The theorem follows as a repeated application of this process.

THEOREM 28. *Let G and G' be dual graphs, and let $\alpha(ab)$, $\alpha'(a'b')$ be two corresponding arcs. Form G_1 from G by dropping out the arc $\alpha(ab)$, and form G_1' from G' by dropping out the arc $\alpha'(a'b')$, and letting the vertices a' and b' coalesce if they are not already the same vertex. Then G_1 and G_1' are duals, preserving the correspondence between their arcs.*

Let H_1 be any subgraph of G_1 and let H_1' be the complement of the corresponding subgraph of G_1'.

Case 1. Suppose the vertices a' and b' were distinct in G'. Let H be the subgraph of G identical with H_1. Then

$$n = n_1.$$

Let H' be the complement in G' of the subgraph corresponding to H. Then

$$r' = R' - n.$$

Now H' is the subgraph in G' corresponding to H_1' in G_1', except that H' contains the arc $\alpha'(a'b')$, which is not in H_1'. Thus if we drop out $\alpha'(a'b')$ from H' and let a' and b' coalesce, we form H_1'. In this operation, the number of connected pieces is unchanged, while the number of vertices is decreased by 1. Hence

$$r_1' = r' - 1.$$

As a special case of this equation, if H' contains all the arcs of G', we find

$$R_1' = R' - 1.$$

These equations give

$$r_1' = R_1' - n_1.$$

Thus G_1' is a dual of G_1.

Case 2. Suppose a' and b' are the same vertex in G'. In this case, defining H and H' as before, we form H_1' from H' by dropping out the arc $\alpha'(a'a')$. This leaves the number of vertices and the number of connected pieces unchanged. Thus two of the equations in Case 1 are replaced by the equations

$$r_1' = r_1, \qquad R_1' = R_1.$$

The other equations are as before, so we find again that G_1' is a dual of G_1. The theorem is now proved.

THEOREM 29. *A necessary and sufficient condition that a graph be planar is that it have a dual.*

We shall prove first the necessity of the condition. Given any planar graph G, we map it onto the surface of a sphere. If the nullity of G is N, it divides the sphere into $N+1$ regions. For let us construct G arc by arc. Each time we add an arc joining two separate pieces, the nullity and the number of regions remain the same. Each time we add an arc joining two vertices in the same connected pieces, the nullity and the number of regions are each increased by 1. To begin with, the nullity was 0 and the number of regions was 1. Therefore, at the end, the number of regions is $N+1$.

We construct G' as follows: In each region of the graph G we place a point, a vertex of G'. Therefore G' contains $V' = N+1$ vertices. Crossing each arc of G we place an arc, joining the vertices of G' lying in the two regions the arc of G separates (which may in particular be the same region, in which case this arc of G' is a 1-circuit). The arcs of G and G' are now in (1, 1) correspondence.

G' is the dual of G in the ordinary sense of the word. We must show it is the dual as we have defined the term.

Let us build up G arc by arc, removing the corresponding arc of G' each time we add an arc to G. To begin with, G contains no arcs and G' contains all its arcs, and at the end of the process, G contains all its arcs and G' contains no arcs. We shall show

(1) each time the nullity of G is increased by 1 upon adding an arc, the number of connected pieces in G' is reduced by 1 in removing the corresponding arc, and

(2) each time the nullity of G remains the same, the number of connected pieces in G' remains the same.

To prove (1) we note that the nullity of G is increased by 1 only when the arc we add joins two vertices in the same connected piece. Let ab be such an arc. As a and b were already connected by a chain, this chain together with ab forms a circuit P. Let $a'b'$ be the arc of G' corresponding to ab. Before we removed it, a' and b' were connected. Removing it, however, disconnects them. For suppose there were still a chain C' joining them. As a' and b' are on opposite sides of the circuit P, C' must cross P, by the Jordan Theorem, that is, an arc of C' must cross an arc of P. But we removed this arc of C' when we put in the arc of P it crosses. (1) is now proved.

The total increase in the nullity of G during the process is of course just N. Therefore the increase in the number of connected pieces in G' must be at least N. But G' was originally in at least one connected piece, and is at the end of the process in $V = N + 1$ connected pieces. Thus the increase in the number of connected pieces in G' is just N (hence, in particular, G' is itself connected) and therefore this number increases only when the nullity of G increases, which proves (2).

Let now H be any subgraph of G, let H' be the complement of the corresponding subgraph of G', and let H' include all the vertices of G'. We build up H arc by arc, at the same time removing the corresponding arcs of G'. Thus when H is formed, H' is also formed. By (1) and (2), the increase in the number of connected pieces in forming H' from G' equals the nullity of H, that is,

$$p' - P' = n.$$

But

$$r' = V' - p', \qquad R' = V' - P',$$

as G' and H' contain the same vertices. Therefore

$$r' = R' - n,$$

that is, G' is a dual of G.

To prove the sufficiency of the condition, we must show that if a graph has a dual, it is planar. It is enough to show this for non-separable graphs. For if the separable graph G has a dual, its components have duals, by Theorem 25, hence its components are planar, and hence G is planar, by Theorem 27. This part of the theorem is therefore a consequence of the following theorem:

THEOREM 30. *Let the non-separable graph G have a dual G'. Then we can map G and G' together on the surface of a sphere so that*

(1) *corresponding arcs in G and G' cross each other, and no other pair of arcs cross each other, and*

(2) *inside each region of one graph there is just one vertex of the other graph.*

The theorem is obviously true if G contains a single arc. (The dual of an arc ab is an arc $a'a'$, and the dual of an arc aa is an arc $a'b'$.) We shall assume it to be true if G contains fewer than E arcs, and shall prove it for any graph G containing E arcs. By Theorem 8, each vertex of G is on at least two arcs.

Case 1. G contains a vertex b on but two arcs, ab and bc. As G is non-separable, there is a circuit containing these arcs. Thus dropping out one of them will not alter the rank, while dropping out both reduces the rank by 1. As G' is a dual of G, the arcs corresponding to these two arcs are each of nullity 0, while the two arcs taken together of nullity 1. They are thus of the form $\alpha'(a'b')$, $\beta'(a'b')$, the first corresponding to ab, and the second, to bc.

Form G_1 from G by dropping out the arc bc and letting the vertices b and c coalesce, and form G_1' from G' by dropping out the arc $\beta'(a'b')$. By Theorem 28, G_1 and G_1' are duals, preserving the correspondence between the arcs. (Obviously G_1 is non-separable.) As these graphs contain fewer than E arcs, we can, by hypothesis, map them together on a sphere so that (1) and (2) hold; in particular, $\alpha'(a'b')$ crosses ac. Mark a point on the arc ac of G_1 lying between the vertex c and the point where the arc $\alpha'(a'b')$ of G' crosses it. Let this be the vertex b, dividing the arc ac into the two arcs ab and bc. Draw the arc $\beta'(a'b')$ crossing the arc bc. We have now reconstructed G and G', and they are mapped on a sphere so that (1) and (2) hold.

Case 2. Each vertex of G is on at least three arcs. As then G contains no suspended chain, and G is not a circuit and therefore is of nullity $N > 1$, we can, by Theorem 18, drop out an arc ab so that the resulting graph G_1 is non-separable. G' is non-separable, by Theorem 26, and hence the arc $a'b'$ corresponding to ab in G is not a 1-circuit. Drop it out and let the vertices a', b' coalesce into the vertex a_1', forming the graph G_1'. By Theorem 28, G_1 and G_1' are duals, and thus G_1' also is non-separable.

Consider the arcs of G' on a'. If we drop them out, the resulting graph G'' has a rank one less than that of G'. For if its rank were still less, G'' would be in at least three connected pieces, one of them being the vertex a'. Let c and d be vertices in two other connected pieces of G''. They are joined by no chain in G'', and hence every chain joining them in G' must pass through a', which contradicts Theorem 6. If we put back any arc, the rank is brought back to its original value, as a' is then joined to the rest of the graph. Hence, G' being a dual of G, the arcs of G corresponding to these arcs are together of nullity 1, while dropping out one of them reduces the nullity to 0. Therefore, by Theorem 9, these arcs form a circuit P. One of these arcs is the arc ab. The remaining arcs form a chain C. Similarly, the arcs of G corresponding to the arcs of G' on b' form a circuit Q, and this circuit minus the arc ab forms a chain D. C and D have the vertices a and b as end vertices. Also, the arcs of G_1 corresponding to the arcs of G_1' on a_1' form a circuit R. These arcs of G_1' are the arcs of G' on either a' or b', except for the arc $a'b'$ we dropped out. Thus the arcs of G_1 forming the circuit R are the arcs of the chains C and D.

As G_1 and G_1' contain fewer than E arcs, we can map them together on a sphere so that properties (1) and (2) hold. a_1' lies on one side of the circuit R, which we call the inside. Each arc of R is crossed by an arc on a_1', and thus there are no other arcs of G_1' crossing R. There is no part of G_1' lying inside R other than a_1', for it could have only this vertex in common with the rest of G_1', and G_1' would be separable. Also, there is no part of G_1 lying inside R, for any arc would have to be crossed by an arc of G_1', and any vertex would have to be joined to the rest of G_1 by an arc, as G_1 is non-separable.

Let us now replace a_1' by the two vertices a' and b', and let those arcs abutting on a_1' that were formerly on a' be now on a', and those formerly on b', now on b'. As the first set of arcs all cross the chain C, and the second set all cross the chain D, we can do this in such a way that no two of the arcs cross each other. We may now join a and b by the arc ab, crossing none of these arcs. This divides the inside of R into two parts, in one of which a' lies, and in the other of which b' lies. We may therefore join a' and b' by the arc $a'b'$, crossing the arc ab. G and G' are now reconstructed, and are mapped on the sphere as required. This completes the proof of the theorem, and therefore of Theorem 29.

*   *   *   *   *

The paper concluded with a statement of Kuratowski's theorem and a proof that neither K_5 nor $K_{3,3}$ has a dual in Whitney's sense. From this it follows that Kuratowski's characterization can be used to give an alternative proof of the fact that a graph which has a Whitney dual must be planar. In a later paper [11], published in 1933, Whitney succeeded in reversing the argument. He proved that a graph which does not contain either K_5 or $K_{3,3}$ must have a Whitney dual, and thereby obtained a new, less topological, proof of Kuratowski's theorem.

The linking thread of this chapter has been Kirchhoff's idea of a fundamental set of circuits, conceived by him for a very practical purpose. Listing invented the name 'cyclomatic number' and Poincaré and Veblen gave it algebraic form as the nullity of a matrix. Then Whitney used the complementary concepts of rank and nullity as the foundation for his definition of duality. In 1935, he carried the process a step further, motivated by the analogy between the structure of the edge-set of a graph and the structure of a vector space. This analogy is exemplified in the following table:

subsets of edge-set $\longleftrightarrow$ subsets of vector space
circuit $\longleftrightarrow$ minimal dependent set
spanning tree $\longleftrightarrow$ basis
rank $\longleftrightarrow$ dimension of linear span.

In a paper 'On the abstract properties of linear dependence' [12], Whitney axiomatized the properties of a rank function. The resulting structure, called a **matroid,** is a generalization of both graphs and vector spaces. It has proved useful, not only in graph theory, but also in several other branches of mathematics.

References

1. KIRCHHOFF, G. R. Über den Durchgang eines electrischen Stromes durch eine Ebene unbesondere durch eine Kreisformige. *Ann. Phys. Chem.* **64** (1845), 497–514 = *Ges. Abh.*, 1–17.
2. POINCARÉ, H. Second complément à l'Analysis Situs. *Proc. London Math. Soc.* **32** (1900), 277–308 = *Oeuvres*, Vol. 6, 338–372.
3. DEHN, M. and HEEGAARD, P. Analysis situs. *Encyklopädie der mathematischen Wissenschaften* IIIAB3 (1907), 153–220.
4. VEBLEN, O. Theory of plane curves in non-metrical analysis situs. *Trans. Amer. Math. Soc.* **6** (1905), 83–98.
5. DUDENEY, H. E. *Amusements in mathematics.* Nelson, London, 1917.
6. LOYD, S., JNR. *Sam Loyd and his puzzles.* Barse, New York, 1928.
7. KURATOWSKI, K. Proceedings of the *Colloquio Internazionali sulle Teorie Combinatorie.* Rome, 1973.
8. [Abstract 179]. *Bull. Amer. Math. Soc.* **36** (1930), 214.
9. KÖNIG D. Über Graphen und ihre Anwendung auf Determinantentheorie und Mengenlehre. *Math. Ann.* **77** (1916), 453–465. [10D]
10. WHITNEY, H. Non-separable and planar graphs. *Proc. Nat. Acad. Sci. U.S.A.* **17** (1931), 125–127.
11. WHITNEY, H. Planar graphs. *Fund. Math.* **21** (1933), 73–84.
12. WHITNEY, H. On the abstract properties of linear dependence. *Amer. J. Math.* **57** (1935), 509–533.

9

The Four-Colour Problem—to 1936

G. D. BIRKHOFF (1884–1944)

IN THIS chapter we shall continue our account of the unsuccessful search for a solution of the four-colour problem. The story is set mainly in the United States of America, where several leading mathematicians took an interest in the problem, and made important (though not decisive) contributions. A recurring theme is the idea of formulating the conjecture in new ways, so that it might become more amenable to powerful techniques from other branches of mathematics.

The first attempts to reformulate the problem

The first alternative form of the four-colour conjecture was Tait's 'theorem' on colouring the edges of a trivalent planar map. We described this formulation in Chapter 6, and its implications will be fully discussed in the following chapter, since it is a special case of the general problem treated therein.

In 1898 another formulation was published by Heawood, who at that time had probably the clearest insight into the intricacies of the question. His paper 'On the four-colour map problem' [1] began with the observation that if the number of edges bounding each region of a trivalent map is divisible by three, then the regions may be coloured with four colours. He went on to generalize this result in the following way. Suppose that we can assign to each vertex of a trivalent map one of the numbers $+1$, -1, in such a way that the sum of the numbers bordering each region is divisible by three; then the map is four-colourable. Conversely, if the map

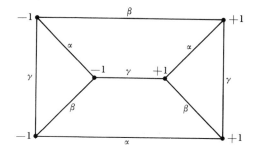

FIG. 9.1.

is four-colourable, then the numbers may be assigned in the prescribed manner. (The special case when the number $+1$ is assigned to each vertex corresponds to the situation mentioned above, that is, when the number of edges bounding each region is divisible by three.)

The proof of the equivalence is very simple. A four-colouring of the regions of a trivalent map is equivalent to a three-colouring of its edges, as explained by Tait; if the edge colours are called α, β, and γ, then we assign $+1$ to a vertex where the clockwise order of the colours is $\alpha\beta\gamma$, and -1 to a vertex where the clockwise order is $\alpha\gamma\beta$. Fig. 9.1 shows the assignment corresponding to the colouring of the map given in Fig. 6.16. In general, it turns out that the sum of the numbers on the boundary of each region is always divisible by three, and that the whole procedure is reversible.

It follows that if the vertices of a trivalent map are labelled $v_1, v_2, \ldots, v_n$, then we have a system of congruences of the form

$$x_i + x_j + \ldots + x_k \equiv 0 \pmod 3,$$

where there is exactly one congruence for each region of the map. Each unknown is $+1$ or -1, and x_i occurs in the congruence corresponding to a region R if and only if v_i is on the boundary of R. Furthermore, every unknown appears in exactly three of the congruences, since there are three regions at each vertex. The truth of the four-colour conjecture is equivalent to the assertion that, for every trivalent planar map, the corresponding system of congruences has a solution.

Heawood devoted a great deal of time and energy to the investigation of such systems of congruences. He published papers on the subject in 1898 [1], 1932 [2], 1936 [3], and even later; the last in the series appeared in 1950, when he was nearly 90. These papers are fairly technical and not particularly easy to read, but the approach is patently no more and no less successful than any other yet discovered.

Although mathematicians are now inclined to view with scepticism any purported proof of the four-colour conjecture, it seems that around the turn of the century the conjecture was still commonly regarded as a topological curiosity, and many people thought that some bright young man would construct a proof with a single stroke of genius. Indeed, the story is told [4, pp. 92–93] of how Hermann Minkowski decided, in the middle of one of his lectures, that he was to be the bright young man. He set out to prove the conjecture there and then, but failed to do so in the time allocated for the lecture, and the next lecture, and the next .... Eventually, he had to admit defeat.

The next significant developments were a symptom of the emerging mathematical might of the United States of America. We have already remarked on the part played in the early history of the problem by the mathematicians of Johns Hopkins University (Peirce, Story, and Sylvester), and the continuing American interest is evident from an address Peirce [5] gave on the subject to the National Academy of Sciences in November 1899. Another mathematician who worked on the problem around that time was Paul Wernicke, a graduate of Göttingen who became professor of mathematics at the University of Kentucky. He read a paper 'On the solution of the map-color problem' to the American Mathematical Society at their Summer meeting in Toronto, in 1897, but the published abstract [6] does not indicate clearly what he said. In 1904 Wernicke published a second paper [7], which foreshadowed the study of the 'reducibility' of maps, a subject we shall discuss in the following section.

It was not until about 1912 that American mathematicians became seriously involved in the four-colour problem, and the credit for this involvement is due primarily to Veblen. His general interest in topology, and his knowledge of the new algebraic methods developed by Poincaré, led naturally to an attack on the four-colour conjecture. In his paper [9A] on the problem, which was read to the American Mathematical Society on 27 April 1912, he employed ideas from finite geometry and incidence matrices over a finite field. At the end of the paper he explained how his formulation may be regarded as a generalization of the system of Heawood congruences.

9A

O. VEBLEN
AN APPLICATION OF MODULAR EQUATIONS IN ANALYSIS SITUS

Annals of Mathematics (2) **14** (1912–3), 86–94.

1. By a map we mean a set of α_2 simply connected regions (countries) covering the surface of a sphere and bounded by α_1 simple arcs (edges) joining α_0 distinct points (vertices). No two regions have a point in common, no two arcs intersect.

For all known maps it is possible to assign to each region one of four colors in such a way that any two regions having an edge in common are differently colored. Whether or not this is true for all maps is still unknown in spite of the investigations of a considerable number of mathematicians. It is therefore perhaps not without interest to show how the problem can be stated in terms of linear equations in a finite field.

These equations turn out to be of service in describing the elementary properties of the map. In particular they supply us with an easy proof of Euler's formula. We shall first outline the discussion of these equations as they arise in the field of integers reduced modulo two and then show how they connect with the four-color problem when the field is extended to include certain Galois imaginaries.

2. A map can be fully described by means of two matrices. To do this, the vertices are numbered in an arbitrary way from 1 to α_0, the edges from 1 to α_1, and the countries from 1 to α_2. In the first matrix the rows correspond to the vertices and the columns to the edges. A "1" appears as the element of the ith row and jth column if the ith vertex is on the jth edge; and a "0" appears as the element of the ith row and jth column if the ith vertex is not on the jth edge. We shall denote this matrix by A; it has α_0 rows and α_1 columns. In the second matrix the rows correspond to the edges and the columns to the countries. The element of the ith row and jth column is 1 if the ith edge is on the jth region and "0" if not. We shall call this matrix B; it has α_1 rows and α_2 columns.

For the map obtained by projecting an inscribed tetrahedron from one of its interior points to the surface of a sphere the matrices A and B are respectively (cf. Fig. [9.2]):

$$A = \begin{pmatrix} 1 & 1 & 1 & 0 & 0 & 0 \\ 0 & 1 & 0 & 1 & 1 & 0 \\ 1 & 0 & 0 & 1 & 0 & 1 \\ 0 & 0 & 1 & 0 & 1 & 1 \end{pmatrix}, \quad B = \begin{pmatrix} 0 & 1 & 0 & 1 \\ 0 & 0 & 1 & 1 \\ 0 & 1 & 1 & 0 \\ 1 & 0 & 0 & 1 \\ 1 & 0 & 1 & 0 \\ 1 & 1 & 0 & 0 \end{pmatrix}.$$

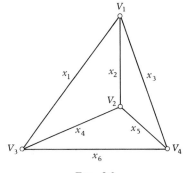

FIG. 9.2.

3. With these matrices may be associated four sets of linear homogeneous equations. In each case the variables and coefficients are regarded as integers

reduced modulo two. In other words, let us add according to the rules $1+1=0$, $1+0=1$, $0+1=1$, $0+0=0$; and multiply according to the rules $1\times1=1$, $1\times0=0$, $0\times1=1$, $0\times0=0$. All the formal laws of elementary algebra are satisfied by this field.

In the first set the equations correspond to the rows of the matrix A. There is one variable for each edge of the map and one equation,

$$(1) \qquad\qquad x_a + x_b + x_c + \cdots = 0$$

for each vertex, the variables in the equation representing the edges which meet at the corresponding vertex. A solution of this system of equations represents a way of labelling the edges of the map with 0's and 1's so that there shall be an even number of 1's on the edges at each vertex. *The edges labelled with 1's in this manner form a number of closed circuits no two of which have an edge in common.* For let us start with an arbitrary edge labelled 1 and describe a path among the edges labelled 1. Whenever there is an edge by which this path approaches a vertex, since the number of 1-edges at this vertex is even, there is a 1-edge by which the path can go away. Hence the path may be continued till it intersects itself. A portion of the path then forms a closed circuit. If this be removed there are still an even number of 1-edges at each vertex. Another circuit may be removed and so on till all the 1-edges are accounted for.

If $(x_1, x_2, \cdots, x_a)$ and $(x_1', x_2', \cdots, x_a')$ are solutions of the equations (1) it is clear that $(x_1+x_1', x_2+x_2', \cdots, x_{a_1}+x_{a_2}')$ is also a solution. The boundary of each of the α_2 countries of the map is represented by a solution in which each edge of the boundary is marked with a 1 and each other edge with a 0. One such solution is supplied by each column of the matrix B. We shall call these the *fundamental solutions*. The solution representing any circuit whatever may be expressed linearly in terms of these α_2 fundamental solutions. In fact, the circuit divides the surface of the sphere into two parts, and the solution representing the circuit is expressible as the sum of the solutions corresponding to the countries in one of these parts.

The sum of the α_2 fundamental solutions is $(0, 0, \cdots, 0)$ because each edge appears on the boundary of two and only two countries. These solutions cannot be subject to any other linear homogeneous relation because the coefficients of such a relation could be only 0 or 1, and hence the relation would merely state that the sum of a certain subset of solutions would be $(0, 0, \cdots, 0)$. This is impossible because any subset of the α_2 countries has at least one country with an edge not on any other country of the subset. Hence *the number of linearly independent solutions of the equations* (1) *is* α_2-1, *and the total number of solutions is* 2^{α_2-1}.

4. A second set of equations is determined by the columns of the first matrix. In these equations the variables correspond to the vertices of the map and there is one equation

$$(2) \qquad\qquad v_a + v_b = 0$$

for each edge, v_a and v_b being the vertices at the ends of the edge. The only possible solutions are such that all the variables are equal. For if v_a is given, v_b must be equal to v_a; if v_c is connected with v_b by an edge, v_c is also equal to v_a, and so on. Since there is a path along the edges joining any vertex to any other it follows by this argument that all the variables are equal to v_a. Hence the only solutions of the equations are $(0, 0, \cdots, 0)$ and $(1, 1, \cdots, 1)$. There are α_0

variables. Hence the number of the α_1 equations which are linearly independent must be $\alpha_0 - 1$. *Hence the rank of the matrix A is* $\alpha_0 - 1$.

5. Let us return a moment to the first set of equations. Here there were α_0 equations among α_1 variables and there were $\alpha_2 - 1$ solutions. The rank of the matrix A has just been seen to be $\alpha_0 - 1$ so that the number of linearly independent equations is $\alpha_0 - 1$. Hence

$$\alpha_1 - (\alpha_0 - 1) = \alpha_2 - 1,$$

or

$$\alpha_0 - \alpha_1 + \alpha_2 = 2,$$

which is the well-known Euler's formula.

6. The third set of equations may be read from the rows of the matrix B. The variables correspond to the countries and for each edge there is an equation of the form

$$(3) \qquad\qquad\qquad y_a + y_b = 0,$$

where y_a and y_b correspond to the countries meeting in the edge in question. These equations are entirely analogous to the equations (2). In fact if a point (the capital of the country.....) be introduced in each region and the points in abutting regions be joined by non intersecting arcs, there is obtained a map (of α_2 points, α_1 edges and α_0 regions) dual to the first map and interchanging the rôles of the matrices A and B. The only solutions of the equations (3) are, by precisely the argument used for the equations (2), $(0, 0, \cdots, 0)$ and $(1, 1, \cdots, 1)$. Hence the rank of B is $\alpha_2 - 1$.

7. The fourth set of equations correspond to the columns of the matrix B. The variables correspond to the edges of the map and for each country there is an equation of the form

$$(4) \qquad\qquad\qquad e_a + e_b + \cdots = 0,$$

where $e_a, e_b, \cdots$ are the edges of the region. Just as in the case of the equations (1), a solution represents a system of circuits among the regions of the map. A circuit is a set of distinct countries $r_1, r_2, \cdots, r_n$ and distinct edges $e_1, e_2, \cdots, e_n$ such that r_1 and r_2 meet along e_1, r_2 and r_3 meet along e_2, $\cdots$, r_n and r_1 meet along e_n. A circuit is *simple* if no subset of its edges and regions form a circuit. To distinguish a simple circuit composed of edges and regions from an ordinary circuit which is composed of edges and vertices we may call the former a *cycle*. The set of all edges and regions meeting at a vertex form a cycle which we shall call *fundamental*. The solutions which correspond to fundamental cycles are given by the rows of matrix A of which $\alpha_0 - 1$ are linearly independent. Since the rank of B is $\alpha_2 - 1$ and the number of variables in the equations (4) is α_1, the number of linearly independent solutions is $\alpha_1 - \alpha_2 + 1$, which by Euler's formula is $\alpha_0 - 1$. Hence the α_0 fundamental cycles furnish a set of solutions of (4) in terms of which all the solutions are expressible; and these fundamental solutions satisfy one linear homogeneous relation.

8. Turning to the four color problem, let us suppose the field $GF(2)$, consisting of 0 and 1 combined modulo two, to be extended by the Galois imaginaries satisfying the relations $i^2 + i + 1 = 0$. The extended field $GF(2^2)$ has four elements, 0, 1, i, $i + 1$ which we may use to denote the four colors. Two elements α, β of this field are equal if and only if $\alpha + \beta = 0$. *Hence a solution of the four color problem consists in finding a set of values* $(y_1, y_2, \cdots, y_{a_2})$ *which satisfies none of the equations* (3) *corresponding to the rows of the matrix B*.

9. The set of values $(y_1, y_2, \cdots, y_{\alpha_2})$ may be regarded as a point in a finite projective space of $\alpha_2 - 1$ dimensions [8] provided we exclude the set $(0, 0, 0, \cdots, 0)$ and regard $(ky_1, ky_2, \cdots ky_{\alpha_2})$ as the same as $(y_1, y_2, \cdots, y_{\alpha_2})$. Each of the equations (3) then represents an $(\alpha_2 - 2)$-space. If the variables y_i range only over the $GF(2)$ there will be $2^{\alpha_2} - 1$ points in the $(\alpha_2 - 1)$ space. If the variables range over the $GF(2^2)$ there will be $(4^{\alpha_2} - 1)/3$ such points. The first space is included in the second and the points of the second space not included in the first may be regarded as *imaginary* with respect to the first space.

In general, there can be no real point which satisfies none of the equations (3) for such a point would represent a coloring of the map by two colors, which is impossible whenever an odd number of regions meet at any vertex. Hence, in general, every real point lies on at least one of the $(\alpha_2 - 2)$-spaces.

An imaginary point can be written in the form $(y_1 + iy_1', y_2 + iy_2', \cdots, y_{\alpha_2} + iy_{\alpha_2}')$ where $(y_1, y_2, \cdots, y_{\alpha_2})$ and $(y_1', y_2', \cdots, y_{\alpha_2}')$ are real. Hence every imaginary point is on a real line. No imaginary point can be on two real lines because two such lines if they intersect at all have a real point in common. But if $(y_1 + iy_1', y_2 + iy_2', \cdots, y_{\alpha_2} + iy_{\alpha_2}')$ satisfies one of the equations (3), so must $(y_1, y_2, \cdots, y_{\alpha_2})$ and $(y_1', y_2', \cdots, y_{\alpha_2}')$, and conversely. Hence a solution of the four color problem is given *by each real line which does not lie on any of the $(\alpha_2 - 2)$ spaces which are represented by equations (3).*

10. If a point of the real $(\alpha_2 - 1)$-space does not satisfy any of the equations (3) corresponding to the edges in a cycle (cf. §7) the countries in the cycle must be assigned alternately the values 0 and 1. This is impossible in a cycle containing an odd number of regions. Hence *every real point* $(y_1, y_2, \cdots, y_{\alpha_2})$ *lies in at least one $(\alpha_2 - 2)$-space corresponding to an edge of each odd cycle in the map.*

If the equations corresponding to the edges in a cycle be added, each variable enters twice in the sum. Hence the sum vanishes. In other words, the equations corresponding to the edges in the cycle satisfy a linear relation. If any subset of these equations be added it is clear from the definition of a cycle that there is at least one variable which enters only once in the sum, and hence the sum does not vanish. Hence the equations corresponding to the edges in a cycle satisfy only one linear relation. A set of n $(\alpha_2 - 2)$-spaces in an $(\alpha_2 - 1)$-space would in general meet in an $(\alpha_2 - n - 1)$-space. But the $(\alpha_2 - 2)$ spaces corresponding to the edges in a cycle satisfy one linear relation and therefore meet in an $(\alpha_2 - n)$-space. Let us denote by $S_{\alpha_2 - n}^c$ the $(\alpha_2 - n)$-space thus determined by a cycle C_n of n regions.

Any point of $S_{\alpha_2 - n}^c$ must satisfy all the equations corresponding to edges of C_n. Hence, if n is odd, a line joining any point of $S_{\alpha_2 - n}^c$ to any point whatever must lie in at least one of the $(\alpha_2 - 2)$-spaces corresponding to the edges in C_n. Therefore a line which furnishes a solution of the four color problem cannot pass through any point of the $S_{\alpha_2 - n}^c$ corresponding to any odd chain C_n. In order that it be possible to color the map it is necessary that there be at least one point $(y_1, y_2, \cdots, y_{\alpha_2})$ not on any $S_{\alpha_2 - n}^c$ for which n is odd.

This condition is also sufficient. For suppose we have a point $(y_1, y_2, \cdots, y_{\alpha_2})$ not on any $S_{\alpha_2 - n}^c$ corresponding to an odd value of n. The countries are accordingly all labelled 0 or 1, and every cycle consisting entirely of 0's or entirely of 1's contains an even number of countries. The map breaks up into a finite number of connected portions, each of which is entirely composed of 0-countries and entirely bounded by 1-countries or entirely composed of 1-countries and entirely bounded by 0-countries. Consider all the 0-countries of a given connected set. They can be reached from an arbitrary 0-country, r_1, by paths which do not pass through vertices and go only through 0-countries. The paths from r_1 to any other country

r_2 of this set cross always an odd or always an even number of edges; for if not, on cancelling the common edges of two paths from r_1 to r_2 there would remain at least one odd cycle of 0-countries. Let us color white the country r_1 and all countries of the set connected with it which are reached by crossing an even number of edges and let us color black all countries of the set which are reached by crossing an odd number of edges. No two white countries are adjacent nor are any two black countries. Treat all the connected sets of 0-countries in this fashion. Treat all the connected sets of 1-countries similarly with the colors red and yellow. The result is a solution of the four color problem. Hence the existence of a point $(y_1, y_2, \cdots, y_{\alpha_2})$ not on any $S^c_{\alpha_2-n}$ for which n is odd implies a solution of the four color problem.

The four color problem has now been reduced to the following form. In a finite projective space of (α_2-1) dimensions with three points on a line there are a certain number of spaces $S^c_{\alpha_2-n}$ of dimensionality α_2-n, one for each odd cycle C_n. They all have one point in common (§6). *The map can be colored in four colors if and only if there exists a point not on any of these $S^c_{\alpha_2-n}$'s. There are as many distinct ways of coloring the map (aside from permutations of the colors) as there are real lines in the (α_2-1)-space which do not meet any $S^c_{\alpha_2-n}$ (n, odd).*

11. Another set of equations associated with the map problem arises as follows. If y_i and y_j are two variables which appear in the same one of the equations (3), i.e., which correspond to adjacent regions, let us denote $y_i + y_j$ by a new variable x_k. There will be one x_k for each of the equations (3), i.e., for each edge of the map. The condition that none of the equations (3) be satisfied now takes the form

$$(5) \qquad\qquad x_k \neq 0 \quad (k = 1, 2, \cdots, \alpha_1).$$

The set of all x's corresponding to the edges meeting at a vertex of the map is a sum of pairs of y's in which each y appears twice. Hence if the x's meeting at a vertex be $x_a, x_b, x_c, \cdots$, they must satisfy the equation,

$$(6) \qquad\qquad x_a + x_b + x_c + \cdots = 0.$$

These equations are evidently the same as (1), §3. In other words, *it is necessary in order to solve the problem to find a solution of the equations (1) in which none of the variables vanishes.*

This is also sufficient. For any equation of the form (6) in which $x_a, x_b, \cdots$ are the edges which appear in a cycle is linearly dependent on the equations (1). (It is in fact the sum of the equations (1) corresponding to vertices in one of the two parts into which the surface of the sphere is divided by the cycle.) Hence a solution of the equations (1) is such that the sum of the marks on the edges of any cycle is zero. Suppose this solution labels all the edges of the map with marks different from 0. If now the mark 0 be assigned to an arbitrary region and the other regions be marked according to the rule that the sum of the marks of two adjacent countries shall be equal to the mark of the edge separating them, a unique mark is assigned to each country; otherwise the edges of some cycle would have a sum different from zero. Since the marks of no two adjacent countries are alike, this determines a coloring in four colors.

12. It is well known that the four color problem may be reduced to the problem of coloring a map in which three edges meet at each vertex. In this case the equations (1) discussed in §11 have only three [terms] each. ......

From the equations (1) in this form can be derived the set of equations modulo three discovered by Heawood [1]. In any solution of (1) the three values of the variables x_a, x_b, x_c corresponding to the edges which meet at any vertex must be 1, i and i^2 in some order, for this is the only way of satisfying

$$x_a + x_b + x_c = 0$$

by values different from zero. An arbitrary sense on the sphere having been chosen as positive, the three marks on the edges at any vertex follow one another in the positive sense either in the cyclic order 1, i, i^2 or the cyclic order 1, i^2, i. In the first case each mark is obtained from its predecessor by multiplying by i, in the second case by multiplying by i^2. Thus by properly distributing the marks i, and i^2 at the vertices the marks of all edges are determined as soon as the mark on one edge is given. For if the mark α on an edge, e, is given and β is the mark at one of its vertices, v, then the mark on the edge following e in the positive cyclic order at v is $\alpha\beta$ and the mark on the other edge is $\alpha\beta^2$. In order that this process assign a unique mark to each edge, the product of the i's and i^2's multiplied in while describing any closed circuit must be unity.

It is necessary and, in view of the simple connectivity of the sphere, sufficient that this condition be satisfied for the boundary of each country in the map. In view of the identity,

$$i^3 = 1,$$

this means that the sum of the exponents of the i's at the vertices of any country must be divisible by three. Hence if z_1, z_2, $\cdots$, z_k are the exponents at the vertices of any country they must satisfy the relation,

$$z_1 + z_2 + \cdots + z_k = 0 \qquad \text{(mod. 3)}.$$

The α_2 countries give rise to α_2 equations of this form among α_0 variables representing the vertices of the map. These are Heawood's equations. The matrix of the coefficients of the equations is the matrix analogous to A and B representing the incidence relations of the vertices and countries of the map.

To solve the four color problem it is necessary and sufficient to find a solution of these equations in which none of the variables vanish. The variables may be interpreted as coördinates of points in a finite projective space of α_0-dimensions in which there are four points on every line.

About this time (1912), one of Veblen's colleagues at Princeton University was G. D. Birkhoff. He became interested in the four-colour problem, and wrote two papers on it during his stay in Princeton. The papers of Veblen and Birkhoff became the major stimuli for the steady stream of American work on graphs and maps from 1912 onwards.

The volume of the *Annals of Mathematics* which contained Veblen's paper on the four-colour problem also contained the first paper by Birkhoff on the same problem. In this paper he initiated a new, quantitative approach to the problem. For a given positive integer λ and a given map M, he introduced the symbol $P(\lambda)$ to denote the number of ways of colouring the regions of M when λ colours are available, subject to the usual restriction that adjacent regions have different colours. For example, if M is the map shown in Fig. 9.3, then we can colour region Q with

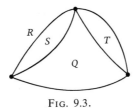

FIG. 9.3.

any one of the λ colours, region R with any of the remaining $\lambda - 1$ colours, and regions S and T with any of the other $\lambda - 2$ colours; consequently, for this map

$$P(\lambda) = \lambda(\lambda - 1)(\lambda - 2)^2 = \lambda^4 - 5\lambda^3 + 8\lambda^2 - 4\lambda.$$

Birkhoff noticed that the function P is always a polynomial in λ, and he derived a general formula for the coefficients of this polynomial. In our extract [9B] we have omitted his complicated determinantal proof of the formula for the coefficients, since a simpler method due to Whitney will be given later in this chapter.

9B

G. D. BIRKHOFF

A DETERMINANT FORMULA FOR THE NUMBER
OF WAYS OF COLORING A MAP

Annals of Mathematics (2) **14** (1912–3), 42–46.

Suppose that a finite set of two-dimensional regions making up a simply or multiply connected closed surface are given, so that these form a *map M*. Each of these regions may be taken to be limited by closed curves, formed by a finite number of continuous *boundary lines* which the region has in common with other regions. The ends of these lines, at which three or more regions meet, are called *vertices* of the map. A *coloring* of the map consists in attributing to each region a color different from that of any region having in common with it a boundary line, but not necessarily different from that of a region meeting it at a vertex.

The following fact will first be proved: *The number of ways of coloring the given map M in λ colors* ($\lambda = 1, 2, \cdots$) *is given by a polynomial $P(\lambda)$ of degree n, where n is the number of regions of the map M.* In fact let $m_i (i = 1, 2, \cdots, n)$ be the number of ways of coloring the map by using exactly i colors *when mere permutations of the colors are disregarded.* With this definition it is clear that

$$m_i \cdot \lambda \cdot (\lambda - 1) \cdots (\lambda - i + 1)$$

represents the number of ways of coloring the given map in exactly i of the λ colors *counting two colorings as distinct when they are obtained by a permutation one from the other*; for, of the i colors used, the first may be chosen in λ ways, the second in $\lambda - 1$ ways, and so on. If λ is less than i the above term reduces to zero.

But the total number of ways of coloring the given map in λ colors is the sum of the number of ways of coloring it with $1, 2, \cdots, n$ of these colors, since no more than n colors can be used. Accordingly the total number of ways is represented by

$$P(\lambda) = m_1 \lambda + m_2 \lambda(\lambda - 1) + \cdots + m_n \lambda(\lambda - 1) \cdots (\lambda - n + 1)$$

for all values of λ. It is clear that in general $m_1 = 0$ inasmuch as for $n > 1$ no map can be colored in a single color, and that $m_n = 1$ since there is only one way of coloring M in n colors if permutation of the colors be disregarded.

In order to proceed to the effective determination of $P(\lambda)$ we consider the total number $\mu - 1$ of ways of forming from the map $M^{(1)} = M$ submaps $M^{(2)}, \cdots, M^{(k)}$ of $n - 1$ regions, $M^{(k+1)}, \cdots, M^{(l)}$ of $n - 2$ regions, and so on to $M^{(\mu)}$ of one region, by successive coalescence of regions adjacent along a boundary line. Such a coalescence may be indicated by the removal of all the common boundary lines of the two regions which coalesce. The maps $M^{(2)}, \cdots, M^{(k)}$ are obtained by one such step, the maps $M^{(k+1)}, \cdots, M^{(l)}$ by two such steps, and so on.

At this point we introduce the symbol (i, k) to denote the number of ways of breaking down the map M in n regions to a submap of i regions by k simple or multiple coalescences, i.e., by picking out maps $M, M^{(\alpha_1)}, \cdots, M^{(\alpha_k)}$, each but the first being a submap of the preceding one, and the last one having i regions. It is apparent that we have $(i, k) = 0$ for $k > n - i$, and that $(i, n - i)$ represents the number of ways of making $n - i$ successive simple coalescences. By definition we take $(n, 0) = 1$ and $(i, 0) = 0$ for $i < n$.

$$*\qquad*\qquad*\qquad*\qquad*$$

The final formula for the number of ways of coloring the given map in λ colors is

$$P(\lambda) = \sum_{i=1}^{n} \lambda^i \sum_{k=0}^{n-1} (-1)^k (i, k).$$

The term in λ^n, corresponding to the identical substitution, has the proper coefficient unity according to our previous convention by which the symbol $(n, 0)$ has the value 1.

As a first example of the formula we take the very simple case of a map of three regions which are adjacent each to each. We will have

$$(2, 1) = 3, \quad (1, 1) = 1, \quad (1, 2) = 3,$$

and

$$P(\lambda) = (3, 0)\lambda^3 - (2, 1)\lambda^2 + [-(1, 1) + (1, 2)]\lambda = \lambda(\lambda - 1)(\lambda - 2).$$

The validity of this formula may be verified at once by noticing that we can color any one of these three regions in λ colors, a second region in the $\lambda - 1$ remaining colors, and the third region in the $\lambda - 2$ colors left after the first two regions are colored.

As a second example we take the case of a map in five regions formed by a ring of three regions bounding an interior and exterior region. In this case the symbols (i, k) which enter have the values

$$(4, 1) = 9, \quad (3, 1) = 22, \quad (3, 2) = 51;$$
$$(2, 1) = 14, \quad (2, 2) = 125, \quad (2, 3) = 150;$$
$$(1, 1) = 1, \quad (1, 2) = 45, \quad (1, 3) = 176, \quad (1, 4) = 150$$

so that

$$P(\lambda) = \lambda^5 - 9\lambda^4 + 29\lambda^3 - 39\lambda^2 + 18\lambda = \lambda(\lambda - 1)(\lambda - 2)(\lambda - 3)^2.$$

In this case also the validity of the formula may be at once verified, for the three regions of the ring must be in three distinct colors, while the interior and exterior regions may be in any fourth color different from these three colors.

Even in this second case the value of the symbols (i, k) is not immediately obtained; and if we have a somewhat more complicated map, for example the map formed by twelve five-sided regions on the sphere, a considerable computation would be necessary to determine $P(\lambda)$ directly from the formula, or from the map itself.

Birkhoff was intrigued by the four-colour problem, and in later years he regretted that he had wasted so much time on it. But he also declared that every great mathematician had at some time attacked the problem, and had, for a while, believed himself successful. From his son, Garrett, we learn that he would ask his wife to prepare suitably complicated maps for him to colour: Mrs. Birkhoff's opinion of this task has not been recorded.

Reducibility

Birkhoff's second paper on map colouring was published in 1913, entitled 'The reducibility of maps' [9]. In this paper he reviewed several ideas due to earlier writers, and welded them into a systematic method of investigation. The line of enquiry which he suggested is still actively pursued to this day.

If there are plane maps which need five colours, then there must be among them a map with the smallest number of regions; such a map is said to be **irreducible**. The basic idea is to obtain more and more restrictive conditions which an irreducible map must satisfy, in the hope that eventually we shall have enough conditions either to construct the map explicitly, or, alternatively, to prove that it cannot exist. Birkhoff began by showing that an irreducible map must be trivalent, and went on to prove that it cannot contain any regions bounded by less than five edges. Suppose, for example, that the map contained a triangle (Fig. 9.4). The removal of the edge ab gives a map with fewer regions which can therefore be coloured with four colours; only three colours (say A, B, C) are used in the vicinity of the vertex c, and so ab can be reinstated and the fourth colour D assigned to the triangular region. In the same way we can rule out a region with four edges, using the method of Kempe-chains as explained in Chapter 6. Further results of this kind were obtained in Wernicke's 1904 paper, which showed that an irreducible map must contain either two adjacent pentagons, or a pentagon adjacent to a hexagon [7].

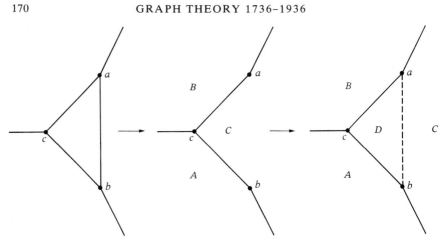

FIG. 9.4.

Birkhoff applied Kempe-chain techniques to obtain results concerning 'rings' of regions. If R is a ring of regions with the maps M_1 and M_2 in its interior and exterior (Fig. 9.5), then a colouring of the maps $M_1 \cup R$ and $M_2 \cup R$ can be combined to give a colouring of the whole map M, provided that these colourings agree on R. By taking M to be an irreducible map, and using a Kempe-chain argument to match a four-colouring of $M_1 \cup R$ and a four-colouring of $M_2 \cup R$, Birkhoff proved that an irreducible map can contain no ring of four regions, and no ring of five regions (unless M_1 is a single pentagon). He also found several other forbidden configurations.

In 1922 Philip Franklin, who had worked with Veblen on analysis situs, made a significant application of the concept of reducibility, in a paper [9C] which showed that the four-colour conjecture is true for all

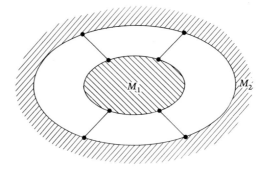

FIG. 9.5.

maps with at most twenty-five regions. His paper illustrates very clearly the types of argument used in proving results about reducibility, and so we have reproduced it in full.

9C

P. FRANKLIN
THE FOUR COLOR PROBLEM

American Journal of Mathematics **44** (1922), 225–236.

1. By a map we shall understand a subdivision of an inversion plane or sphere by means of a finite number of circular arcs into a finite number of regions, which completely cover it. There is no loss of generality in this restriction, as a "map" on any surface of genus zero, with a finite number of regions bounded by simple curves, may be deformed into a map of the type just described. A *side* is a line along which two distinct regions touch each other; a *vertex* is a point which belongs to three or more regions. The problem of coloring a map with a given number of colors (denoted in what follows by A, B, C, etc.) is the association of a color with each region in such a way that any two regions with a side in common are given different colors. Two regions with only a vertex (or a finite number of vertices) in common may of course have the same color.

Whether every map can be colored with four colors is an outstanding question, for while no map has ever been exhibited which could not be colored with four colors, no rigorous demonstration of the possibility for the general case has ever been given. It is known that four colors are necessary to color some maps and that five colors are sufficient to color all maps. If any maps which can not be colored in four colors exist, there must be one such map of a minimum number of regions. We will call such a map an *irreducible* map. It is known that an irreducible map has the following properties [9]:

1. Each vertex belongs to three and only three regions.

2. No group of less than five regions forms a multiply-connected portion of the map. (Consequently there are no two-, three- or four-sided regions and no multiply-connected regions.)

3. No group of five regions forms a multiply-connected portion of the map unless the group consists of the five regions surrounding a pentagon.

4. No edge is surrounded by four pentagons.

5. No region is completely surrounded by pentagons.

6. No even-sided region is completely surrounded by hexagons. Each of these statements amounts to saying that a certain configuration is not possible in an irreducible map. Such a configuration will be called a *reducible* configuration.

In this paper we shall derive a few additional reducible configurations, by means of which it can be shown that the number of regions in an irreducible map is greater than 25. An example is also given of a map of 42 regions which, although colorable in four colors, satisfies all the conditions which have been derived for an irreducible map. It may be taken as showing the extreme lack of generality of the results thus far obtained for this problem.

2. It is interesting to find out some of the properties of a map not containing any region of less than five sides, as this is a property of an irreducible map. Since

our map is drawn on a sphere, the Euler formula (applied to a manifold of genus zero) gives:

(1) $$a_0 - a_1 + a_2 = 2,$$

where a_0, a_1, a_2 are the number of vertices, sides and regions respectively. Also since only three regions touch any one vertex, we have:

(2) $$2a_1 = 3a_0 = \sum_5 \nu A_\nu,$$

where A_ν means the number of regions of ν sides in the map; the last two expressions are each equal to the first since they represent twice the number of sides in the map, counted first with reference to vertices, then with reference to regions. From (1) and (2) we obtain:

(3) $$a_1 = 3(a_2 - 2), \qquad a_0 = 2(a_2 - 2).$$

From (2), (3) and the fact that $a_2 = \sum_5 A_\nu$, we see that

(4) $$6\left(\sum_5 A_\nu - 2\right) = \sum_5 \nu A_\nu.$$

This may be written:

(5) $$A_5 = 12 + \sum_7 (\nu - 6) A_\nu$$

and since the second term on the right is positive, A_5 must be at least 12, and we have the well-known theorem:

Every map containing no triangles or quadrilaterals and having three regions abutting on each vertex contains at least twelve pentagons [6B].

We shall also prove that such a map must contain either:

A pentagon adjacent to two other pentagons,
A pentagon adjacent to a pentagon and to a hexagon, or
A pentagon adjacent to two hexagons.

(That every such map contains either two adjacent pentagons or a pentagon adjacent to a hexagon was proved by P. Wernicke [7].)

For, consider a map with none of these combinations of regions and let us count the number of vertices in the map which belong to a hexagon or a pentagon. We find that the number of vertices contributed by hexagons nowhere in contact with pentagons will be greater than twice the number of such hexagons since each hexagon has six vertices and no vertex belongs to more than three hexagons; pentagons isolated from hexagons or other pentagons will give five vertices each; two pentagons adjacent to each other but to no other pentagons or hexagons give eight vertices together, and hence average four each; while a pentagon adjacent to a hexagon gives over four vertices, since of its five vertices we need only deduct two thirds to account for the two where the hexagon joins it. Thus if none of the three conditions enumerated above existed, the number of vertices would be at least $4A_5 + 2A_6$. That is we would have to have:

(6) $$a_0 \geq 4A_5 + 2A_6.$$

But from (5) and the obvious inequality:

$$(7) \qquad 0 \geq \sum_7 (7-\nu)A_\nu$$

there results:

$$(8) \qquad \sum_7 A_\nu + 12 \leq A_5$$

or

$$(9) \qquad \sum_5 A_\nu + 12 \leq 2A_5 + A_6$$

and since (from (3)):

$$(10) \qquad \sum_5 A_\nu = a_2 = a_0/2 + 2,$$

$$(11) \qquad a_0/2 + 14 \leq 2A_5 + A_6,$$

$$(12) \qquad a_0 + 28 \leq 4A_5 + 2A_6,$$

which contradicts (6) and thus proves the theorem.

The above theorem is not restricted to irreducible maps, but it follows from it that if the configurations there shown to be present were reducible there could not be any irreducible maps and the four-color problem would be solved. While it does not appear to be possible to prove this, there are a number of more complicated configurations which are reducible.

3. To obtain these configurations, and prove their reducibility, we shall need the notion of *chains*, originated by Kempe, and employed by Birkhoff. If a map is colored, or partially colored, a group of regions colored in two colors (say A and B), forming a connected region, and such that each region adjacent to a region of the group is either colored in one of the remaining two colors (C or D) or not yet colored, is said to form a *chain* (an AB chain). Evidently we may obtain a second coloration or partial coloration of the map by interchanging the two colors on a single chain, and unless the map contains only one chain in this pair of colors, the new coloration will differ from the old by more than a mere permutation of the colors of the whole map.

Furthermore since two chains with no color in common, as an AB chain and a CD chain, can not "cross" each other, if we have a closed circuit consisting of an AB chain, or an AB chain and an uncolored region in the case of a partially colored map, it follows that the C and D regions on one side of the closed circuit can not belong to the same CD chain as those on the opposite side of the circuit. Thus in Fig. [9.6], if 1 is an uncolored region, and 2 and 4 are joined by an AB chain, 3 must belong to a CD chain distinct from the one containing 6 and 7. Consequently we may interchange the colors in the CD chain containing 3 without affecting 6 and 7. Since, in most of the applications of this process, we shall only be concerned with the arrangement of the colors about the uncolored region, and the rest of them are unchanged by this operation, we shall briefly refer to this operation as "changing 3 to a D." The value of these operations will be seen in the proofs which follow.

4. We shall now prove that

A side of a hexagon surrounded by this hexagon and three pentagons is a

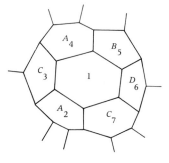

FIG. 9.6.

reducible configuration. For, if it were present in an irreducible map, and we erased the dotted lines as in Fig. [9.7], we would obtain a new map which would contain fewer regions than an irreducible map and hence be colorable. From the way we selected the lines which were erased, regions 1 and 4 would have the same color (say A) while regions 5 and 7 would have a different common color (say B). Of the five essentially distinct colorations, the three cases shown in a, b and c

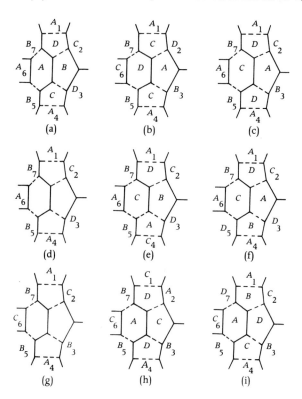

FIG. 9.7.

permit of immediate coloration, as indicated. In the case shown in d, if 5 is joined
to 7 by a BD chain, A may be changed to a C, reducing the problem to case a,
while if 5 is joined to 3 by a BD chain, 4 may be changed to a C, and the map
colored as shown in e. If neither of these chains exist, 5 may be changed to a D,
and the map colored as shown in f. Finally in the case given in g, either a BD
chain joins 7 with 5, and we reduce to c by changing 6 to A; or a BD chain joins
7 with 3, and we color as in h after interchanging A and C in the AC chain
including 1 and 2; or 7 may be changed to a D and we color as in i.

If 5 and 7 had a side in common in our original map, we could not erase the
dotted lines, and still leave a map; but in this case we would have three regions
forming a multiply-connected piece. If 1 and 4 had a side in common, we would
have a group of five regions forming a multiply-connected region of the map, and
not all adjacent to the same pentagon. Hence both these cases are excluded by the
properties of irreducible maps given in the first section.

If a pentagon is in contact with three pentagons, a hexagon, and a fifth region of
any number of sides, either the hexagon is adjacent to this fifth region, in which
case the three adjacent pentagons, with the initial pentagon, completely surround
an edge, or the hexagon is adjacent to two pentagons, and we have a side of the
hexagon completely surrounded by this hexagon and three pentagons. In either
case, it is reducible and we have the result:

A pentagon in contact with three pentagons and a hexagon is a reducible
configuration.

Also if a pentagon is in contact with two pentagons and three hexagons, if the
two pentagons are not adjacent, they are separated by a hexagon, which with the
three pentagons forms a reducible configuration. If the two pentagons are
adjacent, we proceed as follows: We erase the boundaries which are dotted (Fig.
[9.8(a)]) and color the resulting map. If all the regions 1, 2, 3, 4 are not colored in
one color, there are two of them, say 1 and 2, which are separated by a single
region and in different colors, say B and C respectively. We then color 5 with C,
and 6, 7, 8, 9 in turn, which is possible since taking them in this order we shall
never come to a region surrounded by more than three different colors. On the
other hand, if all the regions 1, 2, 3, 4 are colored in the same color, say B, either
there is no BC chain joining all these regions, in which case we can change some
of these regions to C and reduce our problem to the case just discussed, or the
AD chain containing the region 10 is separated from the other regions marked A
and can be changed to a D. The map is then colored as shown in Fig. [9.8(b)].
This proves the theorem:

A pentagon surrounded by two pentagons and three hexagons is a reducible
configuration.

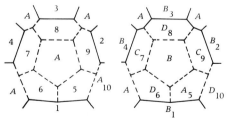

(a)   (b)

FIG. 9.8.

In this proof we have omitted any reference to the case where the dotted lines can not be erased without giving rise to a region which meets itself along one edge. We shall also do this in future cases where, as in this case, it may be excluded by the considerations used for this purpose in the proof of our first theorem.

By a method quite similar to the above, we could easily show that any odd-sided region, completely surrounded by one or more pairs of pentagons, the two of each pair being adjacent, and a number (necessarily odd) of hexagons, is a reducible configuration.

To lead up to a slightly more general theorem, we repeat Birkhoff's proof of the reducibility of an even-sided region surrounded by hexagons, for definiteness stating the proof for a hexagon so surrounded. We erase the dotted lines of Fig. [9.9(a)] and obtain the coloration shown. If all the regions 1, 2, 3, 4, 5, 6 are not colored in one color, there are two of them, say 1 and 2, which are separated by a single region and in different colors, say B and C. We then color 7 with C, and color 8, 9, 10, 11, 12 in turn, which is possible since we shall find each adjacent to regions of three different colors at most, and thus have a fourth with which to color each. If 1, 2, 3, 4, 5, 6 are all in the same color, our map is colored as in Fig. [9.9(b)].

Our generalization is to the case where two adjacent hexagons are replaced by pentagons, and the above method is directly applicable, provided we imagine one of the regions marked A as shrunk to a point. This shows that:

An even-sided region completely surrounded by hexagons and pairs of pentagons, the two of each pair being adjacent, is a reducible configuration.

If a hexagon is surrounded by two hexagons and four pentagons, either the pentagons are grouped so as to come under the theorem just proved, or one of the edges of the hexagon is in contact with three pentagons which we proved above was a reducible configuration. Thus:

A hexagon surrounded by four pentagons and two hexagons is a reducible configuration.

A region of an even number of sides ($2n$) surrounded by $2n-2$ pentagons and two other regions, which are adjacent, is reducible.

To fix the ideas, we state the proof for an octagon. We erase the dotted lines (Fig. [9.10]) and color the resulting map. If 4 is a C, we color 15, 14, 13, 12, 11, 10 in turn which will be possible since each will only be adjacent to regions in at

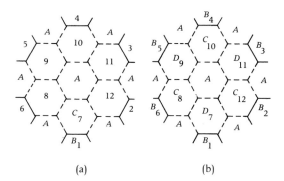

(a)   (b)

FIG. 9.9.

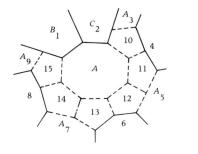

FIG. 9.10.

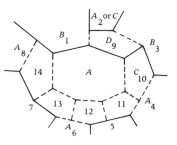

FIG. 9.11.

most three different colors when we come to it. If 4 is a B or D, we color 10 D or B respectively, and color 11 C. If 6 is a C, we color 15, 14, 13, 12 in turn, as before; while if it is B or D we color 12 D or B respectively and 13 C. We then color 14 and 15 D, C; B, D; or B, C; according as 8 is B, C or D.

A region of an odd number of sides $(2n-1)$ *surrounded by* $2n-2$ *pentagons and one other region is reducible.*

We give the proof for a heptagon. After erasing the dotted lines (Fig. [9.11]) we color the new map, as indicated. Reasoning exactly as for the preceding theorem, we show that if 5 is a C the map is colorable, while if it is not C we color 11 and 12, giving 12 the color C. Then we color 13 and 14 D, C; B, D; or B, C; according as 7 is B, C or D.

It follows from our last two theorems that:

An n-gon in contact with $n-1$ *pentagons is reducible.*

5. We will now deduce certain inequalities which must be satisfied by the numbers A_ν if the map is an irreducible one. We have shown that such a map must not contain:

(a) An n-gon in contact with $n-1$ pentagons,
(b) A pentagon in contact with three pentagons and one hexagon,
(c) A pentagon in contact with two pentagons and three hexagons,
(d) A hexagon in contact with four pentagons and two hexagons,

in addition to the six configurations given in the first section, and have also shown that the equation (see (5)):

(13) $$A_5 = 12 + \sum_7 (\nu - 6)A_\nu$$

applies to such a map.

From (a) we know that every region of our map is in contact with at least *two* regions of more than five sides. Hence the number of sides (a side being counted with each of the two regions it separates) of regions with more than five sides must be at least equal to twice the total number of regions. That is:

(14) $$\sum_6 \nu A_\nu \geq 2 \sum_5 A_\nu.$$

To obtain a second inequality from the remaining conditions, we write:

$A_5^0 =$ the number of pentagons in contact with no region of more than six sides,
$A_5^1 =$ the number of pentagons in contact with only one such region,

A_5^2 = the number of pentagons in contact with two or more such regions,
A_6^0 = the number of hexagons in contact with no such regions, and
A_6^1 = the number of hexagons in contact with at least one such region.

From (a) it follows that each region A_5^2, A_6^1 (we thus abbreviate regions of the type counted in A_5^2, A_6^1) as well as those of more than six sides is in contact with at least *two* regions of more than five sides. Also from (a), (b) and (c) it follows that each pentagon A_5^0 is in contact with at least *four* hexagons, and from (a) and (b) that each pentagon A_5^1 is in contact with at least two hexagons in addition to the one region of more than six sides; and therefore to at least *three* regions of more than five sides. Finally from (a) and (d) we see that each hexagon A_6^0 is in contact with at least *three* other hexagons. Thus we have:

(15)
$$\sum_6 \nu A_\nu \geq 4A_5^0 + 3A_5^1 + 2A_5^2 + 3A_6^0 + 2A_6^1 + 2\sum_7 A_\nu.$$

But from the definitions of the regions A_5^1, etc.:

(16)
$$\sum_7 \nu A_\nu \geq A_5^1 + 2A_5^2 + A_6^1.$$

If we add corresponding members of (15) and (16), recollecting that $A_5 = A_5^0 + A_5^1 + A_5^2$ and $A_6 = A_6^0 + A_6^1$, we obtain the result:

(17)
$$\sum_6 \nu A_\nu + \sum_7 \nu A_\nu \geq 4A_5 + 3A_6 + 2\sum_7 A_\nu,$$

which may be written:

(18)
$$2\sum_7 \nu A_\nu \geq 4A_5 - 3A_6 + 2\sum_7 A_\nu.$$

For a map which contains no regions of more than seven sides, we may obtain a somewhat stronger inequality, by using:

(19)
$$4A_6 + 5A_7 \geq 4A_5^0 + 3A_5^1 + 2A_5^2,$$

(20)
$$5A_7 \geq A_5^1 + 2A_5^2,$$

which are analogous to (15) and (16); except that by considering only sides of pentagons in contact with regions of more than five sides we are enabled to use (a) in deriving the left members. They give:

(21)
$$4A_6 + 10A_7 \geq 4A_5,$$

which is only applicable to maps composed entirely of pentagons, hexagons, and heptagons.

By using (13) we may reduce (14) and (18) to the respective forms:

(22)
$$\sum_6 (10 - \nu)A_\nu \geq 24,$$

(23)
$$2\sum_7 (11 - \nu)A_\nu \geq 48 - 3A_6.$$

These two inequalities show that the map we are considering must have at least 25 regions, and if it have only 25, they must be 17 pentagons, 3 hexagons, and 5 heptagons. For if the map contained two hexagons (22) would give:

$$(24) \qquad \sum_{7} (10 - \nu)A_\nu \geq 24 - 4A_6 = 16,$$

which requires at least six regions of more than six sides, and hence by (13) at least 18 pentagons. This would make 26 regions. If the map contained less than two hexagons, the same equation would show that there were more than 27 regions in the map, by a similar argument. Also our map can not have less than five regions of more than six sides, since if the map contained four heptagons, (23) would give:

$$(25) \qquad 3A_6 > 16,$$

and the six hexagons required by (25) together with the 16 pentagons required by (13) would make 26 regions. Furthermore each heptagon less than four will increase the right member of (25) by eight, and hence require at least two additional hexagons in place of the heptagon and pentagon removed. Replacing any of the heptagons by octagons or regions of more than eight sides will strengthen our inequalities, as well as necessitating more pentagons to satisfy (13).

But the map of 17 pentagons, three hexagons, and five heptagons is not irreducible, since it does not satisfy (21). Consequently every irreducible map must contain more than 25 regions and this gives the theorem:

Every map containing 25 or fewer regions can be colored in four colors.

6. The question naturally arises whether 25 is the greatest number for which we can prove such a theorem as the above on the basis of the reductions already described. While an exact answer to this question is lacking, it is evident that the smallest number of regions in a map not containing any of these known reducible configurations is not considerably above 25, as we can construct a map with a small number of regions not containing any of them. Thus in Fig. [9.12] we exhibit a map of 42 regions which satisfies all the properties of irreducible maps given by previous writers as well as those derived in this paper. The map may be formed by constructing a hexagon on each of the 30 edges of a regular

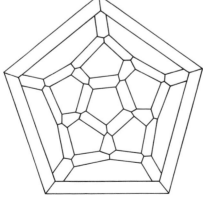

FIG. 9.12.

dodecahedron, in such a way as to leave twelve pentagonal faces, and may be colored by first coloring the pentagons as they would be colored for the dodecahedron.

Franklin's paper was the first in a long line of applications of the concept of reducibility. In 1925 a Belgian mathematician, Alfred Errera, proved that an irreducible map must contain at least thirteen pentagons [10]. This was followed by a paper [11] by C. N. Reynolds, who showed that the number 25 in Franklin's theorem could be increased to 27.

The attractiveness of the method of reducibility is that it enables one to prove positive results on the four-colour problem, by showing that all maps with not more than a certain number of regions are four-colourable. However, such results do not necessarily have any implications for the four-colour conjecture in its stubborn, universal form.

Birkhoff, Whitney, and chromatic polynomials

In the late 1920s, Birkhoff returned to the four-colour conjecture, and, at about the same time, Whitney began to work on a thesis on colouring problems, under Birkhoff's direction. These two events led to some significant developments in the theory of graphs.

Birkhoff took up once again a quantitative approach to the four-colour conjecture, and published a paper [12] in which he proved that, for a map with $n \geqslant 3$ regions

$$P(\lambda) \geqslant \lambda(\lambda - 1)(\lambda - 2)(\lambda - 3)^{n-3},$$

where λ is any positive integer, except 4. Of course, if the result were true for $\lambda = 4$ also, then the truth of the four-colour conjecture would follow.

Whitney began his researches by extending the quantitative approach so that it could be applied to graphs rather than maps. A **colouring of a graph** is an assignment of colours to its vertices, in such a way that vertices which are joined by an edge have different colours; so a colouring of the regions of a map corresponds to a colouring of the dual graph. The general problem of colouring graphs had been mentioned by Kempe at the end of his 1879 paper [6B], but little work had been done on this problem prior to 1930. Whitney used the notation $M(\lambda)$ to denote the number of ways of colouring a given graph when λ colours are available; as in the case of a map, the function M is a polynomial function of λ, and it is now known as the **chromatic polynomial** of the graph.

The first result obtained by Whitney was an expression for the coefficients of the chromatic polynomial similar to the one found by Birkhoff in 1912. Whitney used a general principle, or 'logical expansion', which is sometimes called the 'principle of inclusion and exclusion'; this method has a long history which can be traced back to de Moivre in the early eighteenth century. In a paper [9D] published in 1932, Whitney explained how the principle may be verified, and gave several applications of it in various branches of mathematics. We have reproduced only one such application—his derivation of the formula for the coefficients of the chromatic polynomial.

9D

H. WHITNEY

A LOGICAL EXPANSION IN MATHEMATICS

Bulletin of the American Mathematical Society **38** (1932), 572–579.

1. *Introduction.* Suppose we have a finite set of objects, (for instance, books on a table), each of which either has or has not a certain given property A (say of being red). Let n, or $n(1)$, be the total number of objects, $n(A)$ the number with the property A, and $n(\bar{A})$ the number without the property A (with the property not-A or $\bar{A}$). Then obviously

(1) $$n(\bar{A}) = n - n(A).$$

Similarly, if $n(AB)$ denote the number with both properties A and B, and $n(\bar{A}\,\bar{B})$ the number with neither property, that is, with both properties not-A and not-B, then

(2) $$n(\bar{A}\bar{B}) = n - n(A) - n(B) + n(AB),$$

which is easily seen to be true.

The extension of these formulas to the general case where any number of properties are considered is quite simple, and is well known to logicians. It should be better known to mathematicians also; we give in this paper several applications which show its usefulness. . . . . .

2. *The Logical Expansion.* We prove now the general formula

$$
\begin{aligned}
n(\bar{A}_1\bar{A}_2 \cdots \bar{A}_m) = {}& n - [n(A_1) + n(A_2) + \cdots + n(A_m)] \\
& + [n(A_1A_2) + n(A_1A_3) + \cdots + n(A_{m-1}A_m)] \\
& - [n(A_1A_2A_3) + \cdots] + \cdots + (-1)^m n(A_1A_2 \cdots A_m),
\end{aligned}
$$

(3)

which gives the number of objects *without* certain properties in terms of the numbers of objects *with* various of these properties. The formula is exactly what we should get if we multiplied out the expression in brackets in $n[(1-A_1) \times (1-A_2) \cdots (1-A_m)]$ by ordinary algebra and applied the general formulas

$$n(F+G) = n(F) + n(G), \quad n(-F) = -n(F).$$

We assume the formula holds if there are two properties involved, and shall prove it for the case that three properties are involved. The proof obviously holds if the numbers two and three are replaced by i and $i+1$, and hence the formula is true in general, by mathematical induction.

Consider the objects counted in $n(\bar{A}_1\bar{A}_2)$, that is, those with neither of the properties A_1 and A_2. We wish to know how many of these have the property $\bar{A}_3$. Applying (1) to this set, we have

$$n(\bar{A}_1\bar{A}_2\bar{A}_3) = n(\bar{A}_1\bar{A}_2) - n(\bar{A}_1\bar{A}_2A_3),$$

We know, by hypothesis, that

$$n(\bar{A}_1\bar{A}_2) = n - [n(A_1) + n(A_2)] + n(A_1A_2).$$

To find $n(\bar{A}_1\bar{A}_2A_3)$ we need merely consider those objects with the property A_3, and apply the expansion to this set. Thus

$$n(\bar{A}_1\bar{A}_2A_3) = n(A_3) - [n(A_1A_3) + n(A_2A_3)] + n(A_1A_2A_3).$$

Hence

(4) $n(\bar{A}_1\bar{A}_2\bar{A}_3) = n - [n(A_1) + n(A_2) + n(A_3)]$

$$+ [n(A_1A_2) + n(A_1A_3) + n(A_2A_3)] - n(A_1A_2A_3),$$

as required.

*   *   *   *   *

6. *On the Number of Ways of Coloring a Graph.* Consider any set of objects a, b, c, $\cdots$, f, and any set of pairs of these objects, as ab, bd, $\cdots$, cf. We call the whole collection a *graph*, containing the *vertices* a, b, c, $\cdots$, f, and the *arcs* ab, bd, $\cdots$, cf. It can be visualized simply by letting the vertices be points in space, and letting each arc be a curve joining the two vertices involved.

Suppose we have a fixed number λ of colors at our disposal. Any way of assigning one of these colors to each vertex of the graph in such a way that any two vertices which are joined by an arc are of different colors, will be called an *admissible coloring* of the graph. We wish to find the number $M(\lambda)$ of admissible colorings, using λ or fewer colors.

As a special case of this general problem we have the four-color map problem: To know if we can assign to each region of a map on a sphere one of four colors in such a way that no two regions with a common boundary are of the same color, that is, to see if $M(4) > 0$. Our graph (which is just the dual graph of the map) is constructed by placing a vertex in each region of the map, and joining two vertices by an arc if the corresponding regions have a common boundary. We shall deduce a formula for the number $M(\lambda)$ of ways of coloring a graph due to Birkhoff [9B].

If there are V vertices in the graph G, then there are λ^V possible colorings, formed by giving each vertex in succession any one of the λ colors. Let R be this set of colorings. Let A_{ab} denote those colorings with the property that a and b are of the same color, etc. Then the set of admissible colorings is

$$\bar{A}_{ab}\bar{A}_{bd} \cdots \bar{A}_{cf},$$

and the number of colorings is, if there are E arcs in the graph,

$$M(\lambda) = n(\bar{A}_{ab}\bar{A}_{bd} \cdots \bar{A}_{cf})$$

(11) $$= n - [n(A_{ab}) + n(A_{bd}) + \cdots + n(A_{cf})]$$

$$+ [n(A_{ab}A_{bd}) + \cdots] - \cdots$$

$$+ (-1)^E n(A_{ab}A_{bd} \cdots A_{cf}).$$

With each property A_{ab} is associated an arc ab of G. In the logical expansion, there is a term corresponding to every possible combination of the properties A_{pq}; with this combination we associate the corresponding arcs, forming a *subgraph H* of G. In particular, the first term corresponds to the subgraph containing no arcs, and the last term corresponds to the whole of G. We let H contain all the vertices of G.

Let us evaluate a typical term, such as $n(A_{ab}A_{ad} \cdots A_{ce})$. This is the number of ways of coloring G in λ or fewer colors in such a way that a and b are of the same color, a and d are of the same color, $\cdots$, c and e are of the same color. In the corresponding subgraph H, any two vertices that are joined by an arc must be of the same color, and thus all the vertices in a single connected piece in H are of the same color. If there are p connected pieces in H, the value of this term is therefore λ^p. If there are s arcs in H, the sign of the term is $(-1)^s$. Thus

$$(-1)^s n(A_{ab}A_{ad} \cdots A_{ce}) = (-1)^s \lambda^p.$$

If there are (p, s) (this is Birkhoff's symbol) subgraphs of s arcs in p connected pieces, the corresponding terms contribute to $M(\lambda)$ an amount $(-1)^s(p, s)\lambda^p$. Therefore, summing over all values of p and s, we find the polynomial in λ:

$$(12) \qquad M(\lambda) = \sum_{p,s} (-1)^s(p, s)\lambda^p.$$

Consider a subgraph H of G of s arcs, in p connected pieces. Let us define its rank i and its nullity j by the equations $i = V - p$, $j = s - i = s - V + p$. Then $p = V - i$, $s = i + j$, and putting $(p, s) = m_{V-p,s-V+p} = m_{ij}$, (thus, m_{ij} is the number of subgraphs of G of rank i, nullity j), we have

$$(13) \qquad M(\lambda) = \sum_{i,j} (-1)^{i+j} m_{ij} \lambda^{V-i} = \sum_{i} m_i \lambda^{V-i},$$

$$m_i = \sum_{j} (-1)^{i+j} m_{ij}.$$

7. *The m_i in Terms of the Broken Circuits of G.* We shall find in this section an interpretation of the coefficients of $M(\lambda)$ directly in terms of properties of the graph. Consider for instance the graph G containing the vertices a, b, c, d and the arcs ab, ac, bc, bd, cd, the arcs being given in this definite order. Make a list of the *circuits* in G, naming the arcs of a circuit in the order in which they occur above. Here, the circuits are ab, ac, bc, and bc, bd, cd, and ab, ac, bd, cd. From each circuit we now form the corresponding *broken circuit* by dropping out the last arc of the circuit. The broken circuits here are ab, ac, and bc, bd, and ab, ac, bd. Then *the number* $(-1)^i m_i$ *is the number of subgraphs of G of i arcs which do not contain all the arcs of any broken circuit.*

To show this, we arrange the broken circuits of G in a definite order, where we put a broken circuit P_i before a broken circuit P_j if, in naming the arcs of G one by one in the given order, all the arcs of P_i are named before all those of P_j are named, otherwise, the ordering is immaterial. Suppose there are σ broken circuits, $P_1, P_2, \cdots, P_\sigma$. We now divide the subgraphs of G into $\sigma + 1$ sets (some of which may be empty), putting in the first set, S_1, all those subgraphs containing all the arcs of P_1; in the second, S_2, all those not containing P_1, but containing P_2; in the third, S_3, all those containing neither P_1 nor P_2, but containing P_3; $\cdots$; in the last set, $S_{\sigma+1}$, all those containing none of these broken circuits.

Consider now all the terms in (11) corresponding to the first set of subgraphs. Suppose α_1 is the arc we dropped out of a circuit to form the first broken circuit P_1. To each subgraph in S_1 not containing α_1 corresponds a subgraph in S_1 containing α_1 and conversely, as α_1 is not in P_1. The subgraphs of S_1, and hence the corresponding terms of (11), are thus paired off. But the two terms of each pair cancel. For let H and H' be the two corresponding subgraphs. If H is in p connected pieces, so is H', as the arc α_1 joins two vertices already connected by the broken circuit P_1. The terms each contribute λ^p therefore; but they are of opposite sign, as H' contains one more arc than H.

Consider now the terms corresponding to S_2 (if there are any such). If α_2 is the arc dropped out in forming P_2, α_2 is in neither P_1 nor P_2, on account of the way we have ordered the broken circuits. Thus to each subgraph in S_2 not containing α_2 corresponds a subgraph in S_2 containing α_2, and conversely. The corresponding terms of (11) are thus paired off, and they cancel, exactly as before.

Continuing, we cancel all terms in $S_3, S_4, \cdots, S_\sigma$. We are left only with terms in $S_{\sigma+1}$, that is, those corresponding to subgraphs not containing all the arcs of any broken circuit, and none of these have been canceled.

Consider any such term containing i arcs. The corresponding subgraph H contains no circuit, as it contains no broken circuit. If we build up H arc by arc, each arc we add joins two vertices formerly not connected therefore, and the number of connected pieces is decreased by one each time. Thus the number of connected pieces in H is $V - i$, and the corresponding term contributes $(-1)^i \lambda^{V-i}$ to $M(\lambda)$. If there are l_i such subgraphs, they together contribute an amount $(-1)^i l_i \lambda^{V-i}$. Hence, summing over i, we have $P(\lambda) = \sum_i (-1)^i l_i \lambda^{V-i}$. Comparing with (13), we see that $l_i = (-1)^i m_i$, as required.

EXAMPLES. Let G contain the vertices a, b, c, and the arcs ab, ac, bc. There is one broken circuit: ab, ac. There is one subgraph of no arcs, and $m_0 = 1$. There are three subgraphs of a single arc, and $-m_1 = 3$. There are three subgraphs of two arcs; but one of them contains the broken circuit, so $m_2 = 2$. The subgraph of three arcs contains the broken circuit. Hence, as $V = 3$,

$$M(\lambda) = \lambda^3 - 3\lambda^2 + 2\lambda = \lambda(\lambda - 1)(\lambda - 2).$$

This is easily verified. For we can color a in λ ways; there are $\lambda - 1$ colors left for b; there are now $\lambda - 2$ colors left for c.

Let G be the graph named at the beginning of this section. If a subgraph contains the last broken circuit, it contains the first also, so we can forget the last. We find

$$M(\lambda) = \lambda^4 - 5\lambda^3 + 8\lambda^2 - 4\lambda = \lambda(\lambda - 1)(\lambda - 2)^2,$$

which again is easily verified.

Parts of Whitney's thesis on 'The coloring of graphs' were published as a paper [13], also in 1932. In this paper he continued his study of chromatic polynomials, showing that the numbers m_{ij} (in the notation of [9D]) may be obtained by considering only the non-separable subgraphs, instead of the much larger set of all subgraphs.

Although much of Whitney's other work on graphs (some of which we reviewed in the previous chapter) was indirectly influenced by the four-colour problem, there is one particular direct result of his interest in that problem which deserves special mention. He proved [14] that if we have a trivalent plane map with no loops, then its dual graph must be Hamiltonian. So, in the dual form of the four-colour conjecture—the assertion that every planar graph may be coloured with four colours—it is sufficient to consider only Hamiltonian planar graphs. This is undoubtedly a deep result, as anyone who tries to understand Whitney's paper will agree, and its implications have yet to be exploited fully.

To end this chapter, we return to the work of Birkhoff. In 1934 he published another paper [15] on map-colouring, and, in the first paragraph of his introduction, he outlined the philosophy behind the quantitative approach to the four-colour conjecture:

> In the past, systematic attack upon simple unsolved mathematical problems has invariably led to the discovery of new results of importance as well as to a deeper understanding of the nature of these problems, if not to their actual solution. Perhaps the simplest of all unsolved mathematical problems is that concerned with the validity of the so-called "four-color theorem". The present paper is devoted to the study of the polynomials $P_n(\lambda)$ expressing the number of ways in which a map M_n of n regions covering the sphere can be colored in λ (or fewer) colors; in terms of these polynomials the conjecture of the four-color theorem is that 4 is not a root of any equation $P_n(\lambda) = 0$.

In the rest of his paper Birkhoff discussed the problem of locating the roots of the equation $P_n(\lambda) = 0$. He hoped that the theory of chromatic polynomials might be developed to the point where powerful methods of analytic function theory could be applied.

Of the two methods introduced by Birkhoff, the method of reducibility has yielded more positive results than the method of chromatic polynomials. But there is as yet no really hopeful sign that either method will lead to a final solution of the four-colour problem.

References

1. HEAWOOD, P. J. On the four-colour map theorem. *Quart. J. Pure. Appl. Math.* **29** (1898), 270–285.
2. HEAWOOD, P. J. On extended congruences connected with the four-colour map theorem. *Proc. London Math. Soc.* (2) **33** (1932), 253–286.
3. HEAWOOD, P. J. Failures in congruences connected with the four-colour map theorem. *Proc. London Math. Soc.* (2) **40** (1936), 189–202.
4. REID, C. *Hilbert.* Allen and Unwin, London and Springer, Berlin, 1970.
5. [Report of meeting of Scientific Session, 15 November 1899]. *Rep. Nat. Acad. Sci.* (1899), 12–13.
6. [Report of the Fourth Summer meeting of the Society]. *Bull. Amer. Math. Soc.* **4** (1897–8), 5.

7. WERNICKE, P. Über den kartographischen Vierfarbensatz. *Math. Ann.* **58** (1904), 413–426.

8. VEBLEN, O. and BUSSEY, W. H. Finite projective geometries. *Trans. Amer. Math. Soc.* **7** (1906), 241–259.

9. BIRKHOFF, G. D. The reducibility of maps. *Amer. J. Math.* **35** (1913), 115–128 = *Math. Papers*, Vol. 3, 6–19.

10. ERRERA, A. Une contribution au problème des quatre couleurs. *Bull. Soc. Math. France* **53** (1925), 42–55.

11. REYNOLDS, C. N. On the problem of coloring maps in four colors, I. *Ann. of Math.* (2) **28** (1926–7), 1–15.

12. BIRKHOFF, G. D. On the number of ways of colouring a map. *Proc. Edinb. Math. Soc.* (2) **2** (1930), 83–91 = *Math. Papers*, Vol. 3, 20–28.

13. WHITNEY, H. The coloring of graphs. *Ann. of Math.* (2) **33** (1932), 688–718.

14. WHITNEY, H. A theorem on graphs. *Ann. of Math.* (2) **32** (1931), 378–390.

15. BIRKHOFF, G. D. On the polynomial expressions for the number of ways of coloring a map. *Ann. Scuola Norm. Sup. Pisa* (2) **3** (1934), 85–104 = *Math. Papers*, Vol. 3, 29–47.

10

The Factorization of Graphs

J. P. C. Petersen (1839–1910)

In this, the final chapter of our book, we discuss the problem of 'factorizing' graphs. A special case is Tait's form of the four-colour problem, and we shall begin by describing the tangled web of conjecture and confusion which surrounds his work. The rest of the chapter is devoted to some important contributions to the general theory, due to Julius Petersen and Dénes König. It was König who wrote the first full-length book on graph theory, which was published in 1936 [1].

Regular graphs and their factors

The graphs considered in this chapter are **regular**, that is, every vertex has the same valency. In fact, we shall use the word 'degree', rather than 'valency', since this usage is traditional in studies of factorization.

The basic problem is concerned with splitting a regular graph into other regular graphs which have the same set of vertices. For example, the complete graph K_5 may be split into two pentagons, as in Fig. 10.1. To be precise, we say that the regular graph G has an **r-factor** H if H is a regular graph with degree r, whose vertex-set is the same as that of G and whose edge-set is a subset of that of G. In Fig. 10.1, we depict two 2-factors of K_5.

It is clear that a 1-factor of G consists of a set of disjoint edges, incident with each vertex of G. It follows that a simple graph which has a 1-factor must have an even number of vertices, and, in particular, that K_5 has no 1-factor. The extreme case of factorization occurs when a regular graph of degree k is decomposed into k disjoint 1-factors; in this case we may assign k different colours to its edges in such a way that two edges

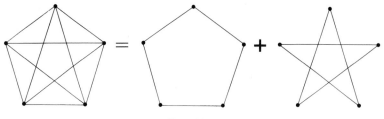

FIG. 10.1.

have the same colour if and only if they belong to the same 1-factor. The result is a colouring of the edges in which each colour occurs just once in the set of edges meeting at any given vertex. This is called an **edge-*k*-colouring** of the graph.

The first reference to questions of this kind occurred in 1880, in two papers [6C], [2] by Tait on map-colouring. In the second of these papers Tait remarked that he had previously given

> ... a theorem which may be stated as follows:-
> *If 2n points be joined by 3n lines, so that three lines, and three only, meet at each point, these lines can be divided (usually in many different ways) into three groups of n each, such that one of each group ends at each of the points.*
> . . . . .
> The difficulty of obtaining a simple proof of this theorem originates in the fact that it is not true without limitation.

The use of the word 'theorem', to describe a result which is 'not true without limitation', is likely to lead to misunderstanding. It did. In its unqualified form, Tait's 'theorem' states that a trivalent graph may be split into three disjoint 1-factors, or equivalently, that it has an edge-3-colouring. Tait himself noticed that the statement is false for trivalent graphs which have a cut vertex, and gave as an example the trivalent graph in Fig. 10.2, which has no edge-3-colouring. But he went on to state his 'theorem' in a form which he believed to be 'universally true':

> *The edges of a polyhedron, which has trihedral summits only, can be divided into three groups, one from each group ending in each summit.*

In this statement there are two implicit conditions, as a consequence of

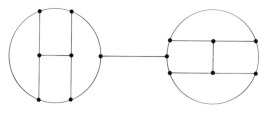

FIG. 10.2.

the fact that the result is now asserted only for the graphs of polyhedra. In fact, a graph may be associated with a convex polyhedron if and only if it is both planar and **3-connected** (that is, at least three vertices must be removed in order to disconnect it). This important result was proved by E. Steinitz in 1922 [3], but its usefulness was not fully appreciated until quite recently.

Tait's paper [2] continued with an account of various methods which he thought would lead to a proof of the restricted 'theorem'. We now realize that the assertion should be regarded as a conjecture, and that it is, in fact, equivalent to the four-colour conjecture; so his remarks should not be taken too seriously.

Tait returned to this topic in an address to the Edinburgh Mathematical Society on 9 November 1883 [4]. He asserted that the statement of his 'theorem' holds, provided that the points lie in a plane and the lines do not cross, and that the figure cannot be disconnected by removing one line. In other words, he recognized that some conditions were necessary, and believed that the result could then be proved. This attitude is understandable (although unjustified), since the result would then follow from Kempe's 'proof' of the four-colour conjecture, which was generally accepted at that time.

In another part of his address, Tait drew attention to a related question. He conjectured that every trivalent polyhedron has a Hamiltonian circuit; and he pointed out that if this were true, then the existence of an edge-3-colouring would follow—since the edges of the Hamiltonian circuit may be coloured α, β, alternately, and the remaining edges coloured γ. He had raised this question in 1880, and had consulted Kirkman on the subject. Kirkman posed the problem (in verse) in the *Educational Times* in 1881, and published a rambling and inconclusive reply [5] to his own query, in which he said that the conjecture 'mocks alike at doubt and proof'. (It was around this time that the two men convinced themselves that Kirkman had been responsible for the idea which led to the Icosian Game.)

In 1946 W. T. Tutte [6] found a trivalent 3-connected planar graph which does not have a Hamiltonian circuit, and thus disproved Tait's conjecture. So Tait was doubly unfortunate: his 'theorem' turned out to be a conjecture, and another conjecture turned out to be false.

The first general discussion of the problem of factorizing graphs was due to Julius Petersen, and appeared in 1891. His interest stemmed from a problem in the theory of invariants, and he set out to apply the graphical notation of Clifford and Sylvester to a question which had arisen in a paper by Hilbert. Petersen's paper is a landmark in graph theory, and the extract [10A] which we give here is only a partial indication of its importance, as we shall explain later.

Two preliminary remarks are necessary. The first is that Petersen always thought of graphs as figures formed by points and lines in a plane, and so extraneous crossings are required in some diagrams. The second remark is that he used the word 'graph' as an English word in a German context, and had it printed in italics throughout. This explains the strange syntax of the original German title.

10A

J. PETERSEN

DIE THEORIE DER REGULÄREN GRAPHS
[The theory of regular graphs]

Acta Mathematica **15** (1891), 193–220.

1. In his proof of the finiteness of the system of invariants associated with a binary form, Hilbert [7] employs a theorem of Gordan concerning a certain class of Diophantine equations. It is a consequence of this theorem that, for a given n, one can construct a finite number of products of the type

$$(x_1 - x_2)^\alpha (x_1 - x_3)^\beta (x_2 - x_3)^\gamma \ldots (x_{n-1} - x_n)^\epsilon,$$

so that all other products of the same type can be built up by multiplying them together. The products are to have the property that the exponents are positive integers (zero included), and that the degree in $x_1, x_2, \ldots, x_n$, is constant for each product. The products so constructed are called *ground-factors;* they correspond to basic solutions of Diophantine equations. For example, if $n = 3$, then one must have $\alpha = \beta = \gamma$ and the only ground-factor is $(x_1 - x_2)(x_2 - x_3)(x_3 - x_1)$. For Hilbert's proof the finiteness of the number of ground-factors is sufficient; for further investigations the actual determination of the expressions is important; this determination is the aim of the following discussion. It makes a remarkable difference whether the degree in each letter (which is equal to the degree of the corresponding invariant) is even or odd. In the first case, which must necessarily occur if the forms have odd order, it turns out that all ground-factors are of the first or second degree (in each letter); in the second case, which is far more complicated, there can be found a ground-factor for which this is not true, for every degree and sufficiently large n. The simplest example has degree three and $n = 10$, and is due to Sylvester, who has undertaken an attack on the question of ground-factors at the same time as myself, and with whom I have often corresponded. Although we have adopted different ways of answering the question, I am nevertheless grateful for his encouragement, without which I would perhaps have been daunted by the great difficulties inherent in each step.

2. One can give the problem a geometrical form by representing $x_1, x_2, \ldots, x_n$ as arbitrary vertices in the plane and indicating each factor $x_m - x_p$ by means of an edge joining x_m and x_p. A product thus gives rise to a figure made up of n vertices joined in such a way that each vertex is incident with the same number of edges. The same two vertices may be joined by several edges. For example, consider

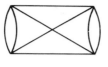

FIG. 10.3.

Figure [10.3] which represents the product

$$(x_1 - x_2)^2 (x_3 - x_4)^2 (x_1 - x_3)(x_2 - x_4)(x_1 - x_4)(x_2 - x_3).$$

English authors have given the name *graph* to such a diagram; I shall retain this name, and say that a graph is regular when each vertex is incident with the same number of edges. Irregular graphs would arise from the consideration of semi-invariants, but they will not be considered here.

By the *order* of a graph I mean the number of its vertices (the order of the binary form) and by its *degree* the number of edges incident with each vertex (the degree of the corresponding invariant). I shall denote by G_α^n, or simply G_α, a graph of order n and degree α. Such a graph can be decomposed into factors when other graphs of the same order but lesser degree can be found, such that superimposing them results in the given graph. A graph which cannot be split in this way is said to be *primitive*. Our problem is the determination of all primitive graphs.

Taking the vertices to be $a, b, c, \ldots$, and denoting the edges by $ab, ac, bc, \ldots$, and the graph by $(ab)^\alpha (bc)^\beta \ldots$, then in this expression each letter occurs equally often, taking account of the exponents.

We shall usually assume that the graph is not made up of graphs of smaller order, that is, that it is not the union of pieces which have no edges connecting them.

Graphs of even degree

3. *A graph of the second degree consists of closed polygons; if all of them have an even number of edges, then the graph can be decomposed into two factors of the first degree; in all other cases the graph is primitive.*

Precisely: we may set out from an arbitrary vertex a and follow an edge ab; there are two edges incident with b; we can continue thence from b to c (let us say); we proceed in this way traversing no edge more than once; so we must eventually return to a, when we have completed a closed polygon; if there are any more edges in the graph, we deal with them in the same way; we stop when all the edges of the graph have been assigned to closed polygons. The two-gons (double edges) can be included in this set.

If from the stated procedure we obtain only one polygon, having an even number of edges, then the graph can be split into two factors of the first degree, since the first, third, fifth, . . . edges constitute a graph of the first degree and the same holds for the remaining edges. When the graph has p polygons, each of which has an even number of edges, then, by taking half the edges of each polygon in the stated manner and using different combinations, the graph may be decomposed in 2^{p-1} different ways. If not all the polygons have an even number of edges, then one can easily see that the graph is primitive.

4. Let us assume that a graph of given degree $\alpha + \beta$ can be decomposed in two different ways, into two graphs of degrees α and β. We can suppose that the edges

of the first graph are blue and the edges of the second red†, so that α blue edges and β red edges are incident with each vertex. The second decomposition can be obtained from the first by switching certain edges between the two factors, that is, by making certain blue edges red and certain red edges blue. If these two systems of edges are represented by symbols $[(ab)^\alpha(bc)^\beta \dots]$, then each letter occurs equally often, since the number of switches involving any letter cannot vary. Supposing, for example, that ab is one of the blue edges switched, then one of the red edges at b, say bc, must be switched, and one of the blue edges at c, say cd, and so on, until one returns to a on a red edge. The edges which are switched must therefore comprise one or more closed polygons, in which the edges are alternately red and blue, and which consequently all have an even number of edges. The second decomposition will be obtained when the colours of the edges of these polygons are all reversed. (Fig. [10.4] and [10.5].)

Such a polygon will be called an *alternating-polygon*, while an *alternating-line* or *alternating-path* will denote an open polygonal line whose sides are alternately red and blue.

Thus we have the following result:

When an arbitrary graph can be decomposed in several ways into two factors (one blue, one red), then each decomposition can be obtained from any other by switching the colours on the edges of certain alternating-polygons.

Similarly, it is quite clear that:

When an arbitrary graph is decomposed into two factors in such a way that an alternating-polygon can be found, then one can obtain a new decomposition by switching the colours on all edges of this alternating-polygon.

5. *Any graph of even degree can be drawn in one continuous stroke, assuming that it is not made up of unconnected pieces.*

In order to demonstrate this theorem, we start from any given vertex a and proceed continuously along the edges in an arbitrary way, except that we are not allowed to traverse any edge more than once. Each time we pass through a vertex, we traverse two of the edges incident with that vertex, while an even number (zero included) of such edges remain to be traversed. When we can proceed no further we must be at a, having used all the edges leading therefrom. We have thus completed a single closed stroke. If all the vertices occurring have been visited as often as possible, then we have traversed the entire graph, for otherwise we have a part of the graph not connected to the rest.

When the route visits a vertex b, but not as often as possible, we break the route at b, and set out in a different way from b, until again the enlarged route is halted at b, all the edges at b having been traversed; in this way we can extend the route until all the edges of the graph are included.

FIG. 10.4.   FIG. 10.5.

† For typographical reasons we shall use continuous lines and dashed lines respectively to represent the blue and red edges.

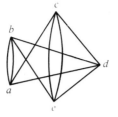

FIG. 10.6.

In the case when the graph is composed of several unconnected pieces, we require as many strokes as there are pieces. If that is not the case, then draw one closed polygon whose edges correspond to the route and which represents the original figure. When the degree is 2α, each letter will appear α times as a vertex of the polygon. We shall call this polygon the *unravelled graph*.

For example, we can consider Fig. [10.6]; we construct the route

$$ab, be, ec, ca, ab, bd, da;$$

breaking it open at d, we insert the route

$$dc, ce, ed,$$

and we can now draw the unravelled graph (Fig. [10.7]). Moreover, one easily sees that the proof of the theorem holds for non-regular graphs, provided that an even number of edges are incident with each vertex. So the theorem holds in particular for any algebraic curve which is not composed of separate pieces, given a suitable convention for infinite branches.

6. *Any graph of the fourth degree can be split into two factors of the second degree.*

Let the graph (Fig. [10.6]) be unravelled in the manner explained above, so that we obtain a polygon (Fig. [10.7]) in which each letter appears twice at a vertex, and which has an even number of edges. Colour the edges of this polygon alternately blue and red, so that a letter labels the intersection of one blue and one red, and each letter occurs twice at the end of a blue edge. Carrying over the colouring of

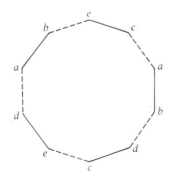

FIG. 10.7.

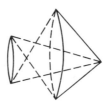

FIG. 10.8.

the edges to the graph, it must follow that each vertex is incident with two red edges and two blue edges (Fig. [10.8]). The graph is thus decomposed into two factors of the second degree.

In the same way, any graph $G_{2\alpha}$ can be split into two factors G_α, provided that the number of edges is even, that is to say, when α and n are not both odd. In that case, the stated decomposition is impossible, for there is no graph of odd order and odd degree. (It is to be taken for granted that the graph is not made up of separate pieces.)

7. Given any graph $G_{2\alpha}$ whose degree is even, we can derive a new graph $G'_{2\alpha}$ by deleting two non-intersecting edges ab and cd and substituting two new edges ac and bd, or ad and bc. If there are several edges ab then we remove only one of them. It is of no significance if one of the replacement edges is already in the graph; it then becomes one of higher multiplicity. I shall call these two *paired graphs*.

When one of two paired graphs can be decomposed into factors of the second degree, then the other can also be so decomposed.

In order to prove this theorem, suppose that $G_{2\alpha}$ is decomposed into factors of the second degree. If ab and cd are in the same factor of degree two, then a new graph of degree two is obtained from it by means of the above construction, while the other factors (which may also contain edges ab and cd) remain unaltered. $G'_{2\alpha}$ is thus decomposed into factors of the second degree. If, on the other hand, ab and cd are in two different factors, then we can take the union of the two factors, which is a graph of the fourth degree. Now, when we carry out the above construction to get $G'_{2\alpha}$, the fourth degree factor becomes another factor of degree four, while all factors of the second degree remain unaltered. However, as described above, we can decompose any graph of degree four into two second degree factors. Thus $G'_{2\alpha}$ is likewise, in this case, decomposed into factors of the second degree.

8. *Any graph may be transformed into any other one of the same order and degree by repeated application of the process described in 7.*

Suppose we wish to transform a given graph into another which contains the second degree factor $ab\ bc\ cd\ de\ ef \ldots ka$. Suppose further that the edges $ab\ bc\ cd$, but not de, occur in the given graph. The edges emanating from d and e cannot all meet at a single vertex; we can therefore find two edges dg and eh, such that g and h are different vertices. We then put de and gh in place of dg and eh, and so introduce de; if there are several edges cd, then it is immaterial if g coincides with c; on the other hand, if cd is simple, we cannot allow g to coincide with c without removing cd, an edge already required. Since the remaining edges emanating from d and e cannot meet in a single vertex, it is however always possible to choose for g a vertex other than c. Thus, we see that it is always possible to insert a new required edge without removing one of those already introduced; we can continue in this way until we have the given second degree factor in the resulting graph. We now remove this factor and obtain a graph, which again can be altered in such a way as to introduce an arbitrary given factor of the second degree. Hence, proceeding in this way, we eventually obtain a graph consisting of arbitrarily chosen factors of the second degree. Any other graph of the same order and degree can likewise be transformed into this, and thence into the original graph.

9. *Every graph of even degree can be decomposed into factors of the second degree.*

In fact, we have seen in 8 that we can construct a sequence of graphs, beginning with one arbitrarily given and ending with one decomposed into factors of the second degree, and such that each two in succession are paired. It follows from the theorem proved in 7 that all graphs in the entire sequence may be decomposed into factors of the second degree.

We have thus attained the first object of our investigations, since we have shown that, when all graphs of the first degree and all primitive graphs of the second degree are given, then all others of even degree can be constructed by superimposition (multiplication). It follows from this that the basic solutions of the Diophantine equations, by means of which the exponents (multiplicities of the edges) are determined, can only be the numbers 0, 1 and 2 in the case under consideration.

We describe the primitive graphs for the first values of n. For $n = 2$ we have a single edge; for $n = 3$, as in 1, a triangle; for $n = 4$, two single edges, representing $(x_1 - x_2)(x_3 - x_4)$ and the two products obtained from this by interchanging symbols. The graphs of odd degree give rise to no new primitive graphs for $n < 10$, as will be explained later. For $n = 5$ there are two kinds of primitive graph, one a pentagon and the other composed of a triangle and a two-gon. For $n = 6$, the graph consisting of two triangles is primitive, while the other graphs of the second degree may be decomposed; and so on.

*   *   *   *   *

As Petersen remarked at the beginning of his paper, the theory of factorization is more difficult for graphs of odd degree than for graphs of even degree. In the part of his paper not included in our extract, he inaugurated the study of graphs of odd degree by proving an important result about trivalent graphs. This result is justifiably famous in its own right, and is related to the four-colour conjecture, so we shall devote our next section to it.

Petersen's theorem on trivalent graphs

A trivalent graph may have only two kinds of non-trivial factor: a 2-factor or a 1-factor. If a 2-factor exists then its complement is a 1-factor, and conversely. If, in addition, there is a 2-factor which itself splits into two 1-factors (that is, if the 2-factor consists entirely of even circuits), then the graph is the union of three disjoint 1-factors. On the

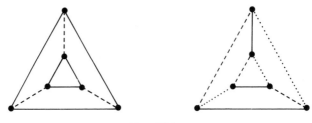

Fig. 10.9.

left of Fig. 10.9, we depict a graph as the union of a 2-factor and a 1-factor; in this case the 2-factor cannot be further decomposed. But the diagram on the right shows the same graph as the union of three 1-factors.

Petersen's famous result, proved in his 1891 paper, is that a regular graph of the third degree always possesses a 1-factor, provided that one simple condition is satisfied. The condition is that the graph should have not more than two leaves, where a **leaf** is a part of the graph which can be disconnected from the rest by removing a single edge. In particular, if the graph arises from a polyhedron, then at least three edges must be removed to disconnect it, so that there are no leaves and Petersen's theorem applies. Petersen's original proof was quite complicated, and we have omitted it in favour of a simpler version [10C] which will be presented at the end of this section.

In 1898 Petersen made another contribution to the theory of trivalent graphs. This came about as a result of a lengthy controversy in the pages of *L'Intermédiaire des Mathématiciens;* as we mentioned at the end of Chapter 6, this debate had begun with some confusion about the status of the four-colour conjecture, and when the subject of Tait's 'theorem' was introduced, it became almost farcical. Petersen [10B] entered the fray better armed than most, since he had a theorem which had actually been proved.

He began by giving an example (which, in his 1891 paper, he had

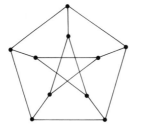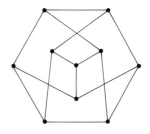

Fig. 10.10.

attributed to Sylvester) of a trivalent graph with no 1-factor (Fig. 10.11). He then stated his major theorem, and noted that Tait's 'theorem' is apparently stronger. However, he went on to exhibit a graph to which the latter result does not apply—in other words, he constructed a trivalent graph, with no leaves, which cannot be split into three disjoint 1-factors. This example has become very well known, and is now called the **Petersen graph** (Fig. 10.15). It has remarkable symmetry properties, which are manifest in the alternative drawings given in Fig. 10.10.

10B

J. PETERSEN
[SUR LE THÉORÈME DE TAIT]
[On the theorem of Tait]

L'Intermédiaire des Mathématiciens **5** (1898), 225–227.

I have developed the theory of regular graphs in a previous memoir [10A]. By the term regular graph I mean a figure consisting of n vertices (n being called the order of the graph) joined by edges in such a manner that d edges are incident with each vertex (d being the degree of the graph). Graphs are closely related to the invariants of binary forms; to each invariant corresponds a graph, and conversely. The order and the degree of the graph are just the order and the degree of the corresponding invariant.

A graph is decomposable when it can be formed by superimposing other graphs of the same order but smaller degree.

There is a very remarkable difference between graphs of even degree and those of odd degree. The former may always be decomposed into graphs of degrees one and two; on the other hand, when the degree is odd it is always possible to find indecomposable graphs, for any such degree. The graph in fig. [10.11] is the simplest example (except, of course, for degree one).

Defining a *leaf* to be part of a graph which can be separated from the rest by removing just one edge, I have proved the following theorem:

An indecomposable graph of the third degree must have at least three leaves.

Thus, a graph of the third degree which has no leaves can be decomposed,

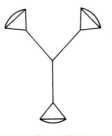

FIG. 10.11.

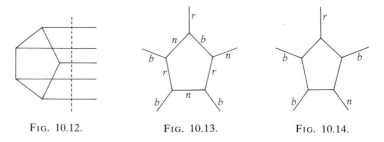

FIG. 10.12.   FIG. 10.13.   FIG. 10.14.

either into three graphs of the first degree, or into one of degree one and one of degree two.

In my terminology, Tait's theorem can be stated as follows:

A graph of the third degree which has no leaves can be decomposed into three graphs of the first degree.

This is, as will be clear, stronger than my theorem and it seems to me impossible that it can be true without the adjunction of extra conditions What is more, as I shall show, I have succeeded in constructing a graph to which Tait's theorem does not apply.

Let us suppose that a graph of the third degree may be decomposed into three others; then one can colour the edges in such a way that each vertex is incident with a blue edge, a red edge and a black edge. (Those points where the edges meet, but which are not vertices, do not enter into this discussion.)

Let us suppress the blue edges; then the resulting graph has degree two and consists of closed polygons. Thus any closed curve (or infinite straight line) must meet it in an even number of points. So, if such a curve cuts x blue edges, y red edges and z black edges, the numbers $x + y$, $x + z$, $y + z$ must all be even. If the number of edges cut is five, then the three numbers can only be 3, 1, 1.

Consider a polygon with five sides, from the vertices of which emanate five edges (fig. [10.12]); there are two possibilities, according as the three blue edges emanate from three consecutive vertices (fig. [10.13]) or from two consecutive vertices and an opposite one (fig. [10.14]).

In the first case a colouring is possible, while in the second case it is easy to see that it is not possible.

We can now easily construct a graph whose decomposition in the manner of Tait's theorem is impossible. Take (fig. [10.15]) a polygon of five sides whose

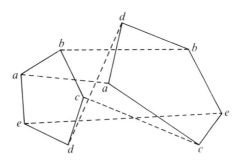

FIG. 10.15.

vertices are, in order, *a, b, c, d, e*, and another whose vertices are assigned the same letters but in the order *a, d, b, e, c*. Join *aa, bb, cc, dd, ee* and the graph is constructed, because three consecutive letters in one polygon cannot be consecutive in the other.

Petersen's remarks drew a reply from É. Goursat [8], who pointed out that, although Petersen had shown that the most general form of Tait's theorem is false, he had not disproved it in the case of polyhedra. (In fact, Petersen's graph is not the graph of a polyhedron, since it is non-planar.) At about the same time, C. de Polignac [9] noted the interesting fact that Petersen's graph is non-Hamiltonian.

In 1899 Petersen published another note [10] on trivalent graphs. He accepted Goursat's comment, and went on to make some observations on the relationship between the four-colour conjecture and the factorization of trivalent planar graphs. At the end of his article he made a surprising statement:

> As I have said, I am not certain of anything, but if I had to wager, I would hold that the theorem of four colours is not correct.

So far we have given no proof of Petersen's main theorem on the existence of a 1-factor in a trivalent graph. The importance of this theorem was recognized by the American school of topologists, and simplified proofs were published by two members of that school—H. R. Brahana in 1917 [11], and Orrin Frink in 1926 [10C]. Unlike its predecessors, Frink's proof involves no counting arguments, but it contains a slight flaw; the published proof of his Theorem I is incomplete, but the necessary repairs are quite straightforward, and may be found in König's book [1, pp. 184–186]. We have included Frink's paper here, since it provides the most direct proof of Petersen's theorem.

10C

O. FRINK
A PROOF OF PETERSEN'S THEOREM

Annals of Mathematics (2) **27** (1925–6), 491–493.

By a regular linear graph of the third degree we mean a network of lines (1-cells) and points (0-cells or vertices) such that there are exactly three 1-cells meeting at every vertex. The number of vertices is called the order of the graph. A regular graph is said to be primitive if it cannot be obtained by superposing two regular graphs of the same order but of lower degree. Sylvester gave an example of a primitive graph of the third degree, but it contained three leaves. Petersen's theorem states that every primitive third degree graph has three or more leaves. By a leaf we mean a portion of the graph which is connected to the rest of the graph only by a single 1-cell, and no proper part of which has this property.

Petersen's proof has been simplified by Brahana [11] and Errera [12]. The present proof is intended to be still simpler and to involve simpler notions. The method used is to consider first graphs without leaves and show that these are factorable (that is, not primitive). The proof is by induction, no shrinking or counting processes being used. A connected regular graph of the third degree without leaves will be called a *simple* graph. It is easy to see that a connected third degree graph is simple when and only when every 1-cell is on a closed circuit.

THEOREM I. *Given a simple graph of order greater than two we can always obtain from it a simple graph of lower order by removing any 1-cell and properly joining the four incident 1-cells in pairs.*

Thus in fig. [10.16] delete x and join 1 to 3 and 2 to 4 obtaining fig. [10.17]. The result will be a simple graph unless the part of the graph containing 1–3 is connected to that containing 2–4 by a single 1-cell or not at all. If this is the case there must be in the original graph a path from 1 to 3 and a path from 2 to 4. For example, if there were no path from 1 to 3 then either 1 or 3 would be the stem of a leaf, or an "isthmus". Thus I say that if we delete x and join 1 to 4 instead and 2 to 3 we obtain a simple graph. To prove this it is sufficient to show that all the closed circuits that were disturbed by the deletion, that is, those through the 1-cells 1, 2, 3 and 4 have been replaced by new ones. The circuits through 1 and 3 have been lengthened, the 1-cell x having been replaced by the path we proved to exist from 2 to 4. The same is true of the circuits through 2 and 4. As for circuits through 1 and 2, if the first deletion disconnected the graph these must also contain 3 and 4 and hence have not been disturbed. If the first deletion resulted in an isthmus however, a circuit through 1 and 2 need not contain 3 and 4 if it contains the isthmus. In this case there are two new circuits, joining 1–3 and 2–4 to the isthmus, one of which will do the work of the original circuit through 1 and 2. The same argument applies in the case of circuits through 3 and 4. The circuits through 1 and 4 and through 2 and 3 have not been disturbed.

It is unfortunate that such a trivial and obvious theorem should require such a long proof. It is to be noted that the argument holds even if a pair of the 1-cells 1,

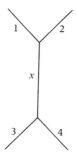

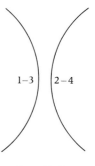

FIG. 10.16.   FIG. 10.17.

2, 3 and 4 are identical or incident at another vertex. If a third degree graph is not primitive, it is said to be factorable, and the first and second degree graphs into which it can be decomposed are called its factors. If the 1-cells of the first degree factor are colored red and those of the second degree factor blue, there will be two blue lines and one red line at every vertex. The graph is then said to be colored. If there exists in a colored graph a closed path consisting of an even number of 1-cells colored alternately red and blue we can interchange colors along this path and obtain a new coloration and hence a new factorization of the graph.

THEOREM II. *Every 1-cell of a colored simple graph is on a closed red–blue path and hence can have its color changed.*

Suppose the theorem were false, then there would exist a colored simple graph of lowest order containing a blue 1-cell, say x, not on any closed red–blue path. Take a blue 1-cell adjacent to x, say y, and delete the red 1-cell incident with its other end joining loose ends according to Theorem I to obtain a simple graph of lower order. In the new graph, which is also colored, there must exist a closed red–blue path through x. But this same path must exist in the original graph, because it is not disturbed by restoring the deleted 1-cell. It is possible that a closed red–blue path containing y might be disturbed by this restoration, but x and y, being adjacent blue 1-cells, cannot be on the same red–blue path. Thus we obtain a contradiction. For a graph of order two Theorem I need not apply, but the conclusion is then obvious.

THEOREM III. *Every simple graph is colorable.*

If the theorem were false there would exist a non-colorable simple graph of lowest order. Lower its order according to Theorem I and color it. Three cases arise. Case I. If both new 1-cells are colored blue, restore the deleted 1-cell colored red. Case II. If one is colored blue and the other red, restore the deleted 1-cell colored blue. In either case we have a coloration of the original graph contrary to assumption. Case III. If both new 1-cells are colored red, change the color of one of them according to Theorem II. If this changes the color of the other also we have Case I, if not, Case II. For a graph of order two the theorem is obvious. This proves the theorem.

THEOREM IV (Petersen's Theorem). *A regular graph of the third degree with fewer than three leaves is colorable.*

We have shown that a simple graph is colorable. A graph with one leaf is impossible. To color one with two leaves, join the leaves by a new 1-cell. The resulting graph is simple and may be colored. If the new 1-cell is colored red, delete it. If it is colored blue, change its color according to Theorem II and then delete it. This restores the original graph, which is now colored.

An alternative view: correspondences

Suppose that we are given a positive integer k and two sets X and Y. A relationship between the members of X and the members of Y, with the property that each x in X is related to exactly k members of Y, and each *y* in *Y* is related to exactly *k* members of *X*, is called a **(k, k)-correspondence** between *X* and *Y*. When $k = 1$ we have the familiar case of a one-to-one correspondence.

Any (k, k)-correspondence may be represented by a graph in the following way. The vertices of the graph represent the members of *X* and *Y*, and two vertices are joined by an edge whenever they are related by the correspondence. For example, Fig. 10.18 represents a $(2, 2)$-correspondence.

A graph representing a (k, k)-correspondence has two special properties. First of all, it is a regular graph of degree *k*; and secondly, the fact that every edge has one end in *X* and one end in *Y* means that the graph is bipartite. The bipartite property can also be expressed by saying that two colours suffice for colouring the vertices of the graph—blue for the vertices of *X*, and red for the vertices of *Y*, for example. An equivalent property is that the graph has no circuits of odd length, since the vertices on any circuit must be coloured alternately blue and red, and so the circuit must contain an even number of vertices and an even number of edges.

If there is a (k, k)-correspondence between two finite sets *X* and *Y*, then we can deduce that *X* and *Y* have the same number of elements; we just count the edges of the representative graph in the following way. Since every edge meets just one vertex in *X* and every vertex is at the end of *k* edges, the total number of edges is $k|X|$. The same argument applied to *Y* shows that the number of edges is $k|Y|$, so that $k|X| = k|Y|$, and hence $|X| = |Y|$. Consequently, the existence of a (k, k)-correspondence between *X* and *Y* implies that there is also a one-to-one correspondence between them. A question which arises naturally is: can we choose the one-to-one correspondence in such a way that, if *x* and *y* are related by it, then they are related in the original (k, k)-correspondence?

In terms of the representative graph, the question asks whether one can find a subset of the edges with the property that every vertex meets just one of them. For example, the graph in Fig. 10.18 has such a set of edges,

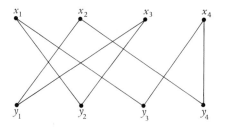

FIG. 10.18.

namely, x_1y_2, x_2y_4, x_3y_1, x_4y_3. So, in the terminology of graph theory, the question becomes: does every bipartite regular graph have a 1-factor?

It was this question which, around the year 1910, inspired the Hungarian mathematician Dénes König to become interested in the theory of graphs. Later [13, p.126], he attributed the idea of the graphical representation of correspondences to his father; and indeed, in a paper [14] published in 1906, the elder König had used a technique which can clearly be expressed in terms of paths in the representative graph.

On 7 April 1914, Dénes König addressed a Congress on Mathematical Philosophy in Paris. He stated (without proof) that every bipartite regular graph has a 1-factor, and deduced as a simple corollary that such a graph may be decomposed into k disjoint 1-factors. The text of his address was not published until 1923 [15], and in the intervening years he had published a proof in both Hungarian [16] and German [10D].

10D

D. KÖNIG

ÜBER GRAPHEN UND IHRE ANWENDUNG AUF DETERMINANTENTHEORIE UND MENGENLEHRE

[On graphs and their applications in determinant theory and set theory]

Mathematische Annalen **77** (1916), 453–465.

The following article deals with problems from. *analysis situs*, the theory of determinants, and set theory. The central notion, by means of which these problems are related to one another, is that of a *graph*. The method of graphs displays, by virtue of its highly geometrical interpretation (which is, indeed, leading to the solution of many recently-discovered questions), the equivalence of apparently remote researches.

§1. **Graphs**

Let a finite number of points be given; then one can choose certain pairs of the points so that one or more (but finitely many) edges join them. A figure constructed in this way we shall generally call a *graph*. A graph is said to be *connected* when one can travel along its edges from any of its points to any other. If the graph is not connected, then it splits in a unique way into connected pieces. From this trivial assertion there arise certain (algebraic) applications of graphs. When the same number of edges leave each point, then the graph is said to be *regular*, and this constant number we shall designate as the *degree* of the graph.

The so-called *bipartite* graphs are of considerable importance; a graph will be so designated when every circuit which can be formed from its edges contains an even number of edges (for example, the graph of the cube).

A graph is bipartite if and only if its points can be divided into two sets so that only points in different sets are joined directly (by means of an edge).

Proof: If the partition into two such sets is possible, then the points belonging to any circuit must *alternate* between one set and the other; the circuit must therefore contain an even number of edges. And conversely, if one arranges the points of any path alternately in one set or the other, then this can only lead to a contradiction when one follows a closed path containing an odd number of points. (This result holds also for disconnected graphs, except that then the partition of points into two sets is not unique).

In this way the "disconnected" and "bipartite" graphs can be defined in a mutually reciprocal way, for one may characterize disconnected graphs by means of the fact that their points can be separated into two sets in such a way that only points in the *same* set can be joined directly to one another.

A collection G_k of edges of a graph G is said to be a *factor of the kth degree* when each point of G is incident with k edges of G_k. A factor of the kth degree is itself a regular graph of the kth degree, its points coinciding with those of the original graph. In particular, a set of edges constitutes a factor of the first degree when each point of G is the endpoint of just one of these edges. If G is regular of the nth degree, then those edges which are *not* contained in a factor of the kth degree form a factor of the $(n-k)$th degree G_{n-k}. In this case one writes, following Petersen,

$$G = G_k G_{n-k}.$$

Similarly, one can define a decomposition of a regular graph into more than than two factors. The sum of the degrees of the factors is equal to the degree of the given graph. In what follows we shall be dealing especially with the decomposition of a regular graph into factors of the first degree.

Every graph of the second degree is made up of simple (non-self-intersecting) circuits. It follows that one can find a factor of the first degree if and only if all the circuits have an even number of edges [10A]. We shall prove the following generalization of this statement:

A) *Every bipartite regular graph possesses a factor of the first degree.*

This theorem is a direct consequence of the following (which is, however, apparently more significant):

B) *Every bipartite regular graph of the kth degree splits into k factors of the first degree.*

This theorem will, in turn, arise as a consequence of the following:

C) *Supposing that each point of a bipartite graph is incident with at most k edges, then one can assign one of k labels to each edge of the graph in such a way that two edges incident with the same point must have different labels.*

Theorem B) is in fact a direct consequence of this, since, in the case of a regular graph of the kth degree, the edges having a given label comprise a factor of the first degree.

For the proof of Theorem C) we shall employ the method of induction, which, although it is inapplicable to Theorem B), leads directly to the objective in Theorem C). If the number of edges is $\leq k$, then the theorem is clearly true. We suppose then that it is correct when this number is $< N$, and prove the result for a graph G with N edges.

If we delete any edge, e say, from G, there results a (similarly bipartite) graph G', to whose edges we can assign k labels in the manner prescribed in Theorem C). We think of these labels as on the edges of G. Suppose that the deleted edge e joins the points A and B; then, *first*, it is possible that every one of the labels at A (that is, on an edge incident with A) also occurs at B. In this case one can assign a label to the edge e, and our goal is attained. We can thus restrict ourselves to the *second* case: a label, say "1", which is missing at B (one such must exist, since at most $k-1$ edges are incident with B, and also with A) occurs at A. In the same way, let "2" be a label not occurring at A, and let AA_1 be the edge with label "1". It may happen that there is an edge A_1A_2 with label "2", and possibly an edge A_2A_3 with label "1", an edge A_3A_4 with label "2", and so on. We continue this "alternating" path $AA_1A_2A_3\ldots$ for as long as possible. Depending only on the choice of the labels "1" and "2", it is uniquely determined. No point can occur twice in this path, for if A_i were the first such point, which had already been passed through once, then three edges with labels "1" and "2" would occur at A_i. Also, one cannot return to A, since "2" certainly does not occur there and "1" has already been used (on the edge AA_1). Finally, one cannot reach B on this path, since this could only happen by an edge labelled "2" ("1" does not occur at B) and then we should have a path from A to B containing an even number of edges; with the restitution of the deleted edge e, this path gives a circuit with an odd number of edges—in contradiction to our hypothesis.

$AA_1A_2\ldots A_r$ is thus a double-point-free open path, not containing the point B, whose edges carry alternately the labels "1" and "2". Interchange the labels "1" and "2" on the edges of this path without altering the labels on the remaining edges. The distribution of labels resulting from this interchange is sufficient for our purpose. One sees this immediately for the first point A, with which no edge labelled "2" was incident originally, for the internal points A_i, and indeed for the endpoint A_r, with which no two edges labelled "1" or "2" are incident since otherwise our path would not end at A_r. We have now arranged that the label "1" no longer occurs at A_1 and so it can be assigned to the deleted edge e.

Consequently Theorem C) is proved, and thence also Theorem A) and B).

Before we pass on to the applications of the foregoing to determinant theory and set theory, we remark that our Theorem B) is related to a well-known problem of *analysis situs:* the problem of *map-colouring.* The hitherto unproved *four-colour theorem* asserts the following: it is possible to assign one of four colours to the countries of a plane map in such a way that two countries with a common boundary have different colours. One can easily see that it is sufficient to consider the case when the boundaries in the map form a regular graph of the third degree, and that then the four-colour theorem is equivalent to Tait's theorem, which states that such a graph splits into three factors of the first degree. Of course, it is not required for this purpose that every graph of the third degree splits into three factors, for a graph may only be regarded as the system of boundaries of a plane map when the following conditions are satisfied: 1) it may be drawn on a plane (or sphere) in such a way that no new crossing points arise, and 2) each one of its edges belongs to some simple circuit. Without these two restrictions, Tait's theorem (which Tait formulated [4] without conditions and claimed to be self-evident) is certainly false, as we can show by means of two examples [10B]. However, if the graph is a bipartite graph then the theorem is true without the two restrictions, according to Theorem B). Indeed, this is well known in the four colour problem, for in this case—as Kempe [6B] has shown—not just four, but even three, colours suffice. In order to make progress in the general case, and indeed by simpler means, we shall require the characterization of graphs satisfying condition 1) by intrinsic (combinatorial) properties. To that end, the Euler polyhedral theorem plays an important role [17].

$$* \quad * \quad * \quad * \quad *$$

König wrote several other papers on the subject of correspondences. In one of them, published in 1926 [18], he and Stephan Valkó extended the results of [10D] to the case of infinite graphs.

There are two other noteworthy theorems which should be mentioned here. Although König's theorem is apparently unrelated to them, it is in fact a consequence of either one.

The first theorem is a result of K. Menger, first published in 1927 in a paper [19] dealing with a problem in analytic topology. When expressed in the terminology of graph theory, it concerns the relationship between the number of disjoint paths joining two given vertices v and w, and the number of vertices which must be removed in order to ensure that v and w lie in different components of the remaining graph.

The second theorem is due to the group-theorist Philip Hall, and was published in 1935 [20]; it is often expressed informally in the following way. Suppose we have a set X of boys and a set Y of girls, and that we know which boys are acquainted with which girls; that is, we are given a correspondence between X and Y, although it does not necessarily have the regular properties of a (k, k)-correspondence. In this situation, an important question is: can we find a subset of girls such that each boy can marry a girl with whom he is acquainted? In other words, we seek a subset Y_1 of Y such that each x in X can be assigned a unique y in Y_1, x and y being related in the given correspondence. Hall's necessary and sufficient condition is simple to state: we can find a suitable subset of the girls if and only if, for each subset S of the boys, there are at least $|S|$ girls acquainted with a boy in S, where $|S|$ denotes the number of boys in S.

Somewhat surprisingly, Menger's theorem and Hall's theorem are equivalent; furthermore, it can be shown that König's theorem may be deduced from either one of them.

The year 1936 marked the two-hundredth anniversary of Euler's article on the problem of the Königsberg bridges, and so it was a fitting coincidence that the first book on graph theory should be published in that year. The publication of this book, written by Dénes König and entitled *Theorie der endlichen und unendlichen Graphen*, signified that, after two hundred years, graph theory had at last become a branch of mathematics in its own right.

References

1. KÖNIG, D. *Theorie der endlichen und unendlichen Graphen*. Akademische Verlagsgesellschaft, Leipzig, 1936. Reprinted: Chelsea, New York, 1950.
2. TAIT, P. G. Note on a theorem in the geometry of position. *Trans. Roy. Soc. Edinb.* **29** (1880), 657–660 = *Sci. Papers*, Vol. 1, 408–411.
3. STEINITZ, E. Polyeder und Raumteilungen. *Encyklopädie der Mathematischen Wissenschaften* IIIAB12 (1922), 1–139.

4. TAIT, P. G. Listing's *Topologie. Phil. Mag.* (5) **17** (1884), 30–46 = *Sci. Papers*, Vol. 2, 85–98.
5. KIRKMAN, T. P. Question 6610, solution by the proposer. *Math. Quest. Solut. Educ. Times* **35** (1881), 112–116.
6. TUTTE, W. T. On Hamiltonian circuits. *J. London Math. Soc.* **21** (1946), 98–101.
7. HILBERT, D. Über die Endlichkeit des Invariantensystems für binäre Grundformen. *Math. Ann.* **33** (1889), 223–226 = *Ges. Abh.*, Vol. 2, 162–164.
8. GOURSAT, É. [Sur le théorème de Tait]. *Interméd. Math.* **5** (1898), 251.
9. DE POLIGNAC, C. Sur le théorème de Tait. *Bull. Soc. Math. France* **27** (1899), 142–145.
10. PETERSEN, J. [Réponse à question 360]. *Interméd. Math.* **6** (1899), 36–38.
11. BRAHANA, H. R. A proof of Petersen's theorem. *Ann. of Math.* (2) **19** (1917–8), 59–63.
12. ERRERA, A. Une démonstration du théorème de Petersen. *Mathesis* **36** (1922), 56–61.
13. KÖNIG, D. Sur les correspondances multivoques des ensembles. *Fund. Math.* **8** (1926), 114–134.
14. KÖNIG, J. Sur la théorie des ensembles. *Compt. Rend. Acad. Sci.* (*Paris*) **143** (1906), 110–112.
15. KÖNIG, D. Sur un problème de la théorie générale des ensembles et la théorie des graphes. *Rev. Métaphys. Morale* **30** (1923), 443–449.
16. KÖNIG, D. Gráfok és alkalmazásuk a determinánsok és halmazok elméletében. *Mat. Természettud. Értesítő* **34** (1916), 104–119.
17. KÖNIG, D. Vonalrendszerek kétoldalú felületeken. *Mat. Természettud. Értesítő* **29** (1911), 112–117.
18. KÖNIG, D. and VALKÓ, S. Über mehrdeutige Abbildungen von Mengen. *Math. Ann.* **95** (1926), 135–138.
19. MENGER, K. Zur allgemeinen Kurventheorie. *Fund. Math.* **10** (1927), 96–115.
20. HALL, P. On representatives of subsets. *J. London Math. Soc.* **10** (1935), 26–30.

PLATE 1

W. R. Hamilton (1805–65)

P. G. Tait (1831–1901)

M.-E.-C. Jordan (1838–1922)

O. Veblen (1880–1960)

Plate 2

W. K. Clifford (1845–79)

F. Guthrie (1831–99)

K. Kuratowski (1896–   )

H. Whitney (1907–   )

Graph Theory since 1936

In this appendix, we shall describe briefly some of the progress made in graph theory since 1936. Our aim is not to give a complete survey of all the noteworthy results which have appeared since that date, but rather to take a few specific areas discussed earlier in this book, and outline the subsequent development of the subject in these areas. The reader who wishes to learn more about recent work in graph theory should consult the specialist monographs listed at the end of this appendix.

Paths and circuits

In Chapters 1 and 2, we saw how Euler and Hierholzer obtained a necessary and sufficient condition for a connected graph to have an Eulerian path, but we remarked that the corresponding problem for a Hamiltonian circuit was unsolved. Since 1936 some progress has been made on the latter problem, in that various *sufficient* conditions for a graph to be Hamiltonian have been found. G. A. Dirac [1] proved in 1952 that *if G is a simple graph with $n(\geqslant 3)$ vertices, and if the valency of each vertex is at least $\frac{1}{2}n$, then G is Hamiltonian.* Similarly, O. Ore [2] proved in 1960 that *if G is a simple graph with $n(\geqslant 3)$ vertices, and if the sum of the valencies of each pair of non-adjacent vertices is at least n, then G is Hamiltonian.* These results were later generalized by L. Pósa [3] and V. Chvátal [4].

Enumeration and chemistry

In Chapters 3 and 4, we described the contributions of Cayley, Redfield, and Pólya, and concluded with a brief mention of Pólya's fundamental paper, published in 1937. Pólya's theorem has been generalized by N. G. de Bruijn [5], and it has been applied to the enumeration of many kinds of graphs. For example, R. W. Robinson has enumerated graphs which have an Eulerian path [6], and trivalent graphs. However, little progress has been made in problems relating to the enumeration of k-colourable graphs, or graphs with a Hamiltonian circuit. For a survey of results on enumeration, the reader is referred to [F].

The use of graph theory in chemistry has grown enormously during recent years. Quite apart from its applications to enumerative chemical problems such as those mentioned in Chapter 4, graph theory has been employed in problems relating to chemical nomenclature and documentation, bonding theory, and the analysis of chemical systems. The interested reader is referred to [7] for further details.

Maps on surfaces

In Chapter 5, we introduced the study of maps drawn in the plane and on more complicated surfaces. This led, in Chapter 7, to a discussion of the chromatic number of an orientable surface of genus g, and we presented Heawood's conjecture, that this number is equal to $[\frac{7}{2}+\frac{1}{2}\sqrt{(1+48g)}]$, for all $g \geqslant 1$. The truth of Heawood's conjecture for all values of g was not finally established until 1967;

several mathematicians contributed to the proof, but most of the credit goes to G. Ringel and J. W. T. Youngs. The proof is fairly complicated, involving the consideration of twelve separate cases, corresponding to the residue classes modulo 12. Full details may be found in Ringel's book [I].

The four-colour problem

In Chapters 6 and 9, we outlined the history of the four-colour problem from De Morgan's letter of 1852 until the mid-1930s. Since then, there has been no really significant progress towards its solution, although many interesting results on the colouring of graphs have been obtained.

We showed in Chapter 9 how the study of reducibility enabled Franklin to prove the four-colour conjecture for all maps with at most 25 countries, and we mentioned that this number had been increased to 27 by Reynolds. More recently, it has been increased to 35 by C. E. Winn [8], to 39 by O. Ore and J. Stemple [9], and to 95 by J. Mayer (in work as yet unpublished).

A different approach to the problem was suggested by H. Hadwiger [10], who conjectured that *every k-colourable graph can be contracted to K_k.* (A **contraction** of a graph is a sequence of operations, each of which consists of deleting an edge and identifying its incident vertices.) Hadwiger's conjecture is known to be true for $k = 2$, 3, 4; its truth for $k = 5$ would imply the four-colour conjecture, and conversely.

There have been some interesting developments in the theory of chromatic polynomials. After Birkhoff's death, his colleague D. C. Lewis wrote a long paper [11] which contained all Birkhoff's work on the subject. In recent years, W. T. Tutte has obtained some intriguing results on the zeros of chromatic polynomials of planar graphs. He has proved [12] that such polynomials tend to have a real zero close to $1 + \tau$, where τ is the 'golden ratio' $\frac{1}{2}(1 + \sqrt{5})$.

In addition, there have been two important results on colouring, although they are only indirectly related to the four-colour problem. The first of these is due to R. L. Brooks [13], who proved that *if G is a connected simple graph which is neither a complete graph nor a circuit graph with an odd number of vertices, then the vertices of G can be coloured with only ρ colours, where ρ is the maximum valency of the graph.* A similar result, concerning the colouring of the edges, is due to V. G. Vizing [14], who proved that *if G is a simple graph with maximum valency ρ, then the edges of G can always be coloured with ρ + 1 colours, in such a way that any two edges with a vertex in common are assigned different colours.*

Planarity

In Chapter 8, we saw how Whitney defined the dual of a graph in terms of the concepts of rank and nullity, and that he proved that a graph is planar if and only if it has a dual. An alternative characterization of planar graphs was given by S. MacLane [15]: *a non-separable graph is planar if and only if it has a set of circuits with the property that each edge of the graph lies in exactly two of the circuits.*

We also mentioned the concept of a 'matroid'. The study of matroids has become an important branch of combinatorial theory, and it has led to the simplification of several topics in graph theory. In 1959 W. T. Tutte [16] proved an important result giving a necessary and sufficient condition (similar to that in Kuratowski's theorem) for a matroid to be one which arises from a graph.

Factorization

In Chapter 10, we defined the notion of an *r*-factor in a graph. In 1947 W. T. Tutte [17] obtained a necessary and sufficient condition for a graph G to have a

1-factor, namely that

$$c_0(S) \leqslant |S|,$$

for every subset S of the vertex-set of G; here, $|S|$ is the number of vertices in S, and $c_0(S)$ denotes the number of those components of the graph resulting from the deletion of the vertices in S, which have an odd number of vertices. Tutte also obtained a similar condition for the existence of an r-factor [18].

References

Papers referred to above.

1. DIRAC, G. A. Some theorems on abstract graphs. *Proc. London Math. Soc. (3)* **2** (1952), 69–81.
2. ORE, O. Note on Hamiltonian circuits. *Amer. Math. Monthly* **67** (1960), 55.
3. PÓSA, L. A theorem concerning Hamilton lines. *Magyar Tud. Akad. Mat. Kutató. Int. Közl.* **7** (1962), 225–226.
4. CHVÁTAL, V. On Hamilton's ideals. *J. Combinatorial Theory (B)* **12** (1972), 163–168.
5. DE BRUIJN, N. G. Generalization of Pólya's fundamental theorem in enumerative combinatorial analysis. *Indagationes Math.* **21** (1959), 59–69.
6. ROBINSON, R. W. Enumeration of Euler graphs. In: *Proof techniques in graph theory*, Academic Press, New York, 1969, pp. 147–153.
7. ROUVRAY, D. H. Graph theory in chemistry. *R. I. C. Reviews* **4** (1971), 173–195.
8. WINN, C. E. On the minimum number of polygons in an irreducible map. *Amer. J. Math.* **62** (1940), 406–416.
9. ORE, O. and STEMPLE, J. Numerical calculations on the four-color problem. *J. Combinatorial Theory* **8** (1970), 65–78.
10. HADWIGER, H. Über eine Klassifikation der Streckenkomplexe. *Vierteljschr. Naturforsch. Ges. Zürich* **88** (1943), 133–142.
11. BIRKHOFF, G. D. and LEWIS, D. C. Chromatic polynomials. *Trans. Amer. Math. Soc.* **60** (1946), 355–451.
12. TUTTE, W. T. On chromatic polynomials and the golden ratio. *J. Combinatorial Theory* **9** (1970), 289–296.
13. BROOKS, R. L. On colouring the nodes of a network. *Proc. Cambridge Phil. Soc.* **37** (1941), 194–197.
14. VIZING, V. G. The chromatic class of a multigraph, *Cybernetics* **1** No. 3 (1965), 32–41.
15. MACLANE, S. A structural characterization of planar combinatorial graphs. *Duke Math. J.* **3** (1937), 460–472.
16. TUTTE, W. T. Matroids and graphs. *Trans. Amer. Math. Soc.* **90** (1959), 527–552.
17. TUTTE, W. T. The factorization of linear graphs. *J. London Math. Soc.* **22** (1947), 107–111.
18. TUTTE, W. T. The factors of graphs. *Canad. J. Math.* **4** (1952), 314–328.

Books: references A-D are general books on graph theory, references E-K are more specialized.

A. BEHZAD, M. and CHARTRAND, G. *Introduction to the theory of graphs*. Allyn and Bacon, Boston, 1971.
B. BERGE, C. *Graphs and hypergraphs*. North-Holland, Amsterdam and American Elsevier, New York, 1973.
C. HARARY, F. *Graph theory*. Addison-Wesley, Reading, Mass. 1969.
D. WILSON, R. J. *Introduction to graph theory*. Oliver and Boyd, Edinburgh and Academic Press, New York, 1972.
E. BIGGS, N. L. *Algebraic graph theory*. Cambridge University Press, London, 1974.
F. HARARY, F. and PALMER, E. M. *Graphical enumeration*. Academic Press, New York, 1973.

G. MOON, J. W. *Counting labelled trees.* Canadian Mathematical Monographs No. 1, Montreal, 1970.

H. ORE, O. *The four-color problem.* Academic Press, New York, 1967.

I. RINGEL, G. *Map color theorem.* Springer-Verlag, Berlin, 1974.

J. TUTTE, W. T. *Connectivity in graphs.* University of Toronto Press, Toronto, 1966.

K. WHITE, A. T. *Graphs, groups and surfaces.* North-Holland, Amsterdam, 1973.

APPENDIX 2

Biographical Notes

THIS appendix contains short biographies of all the authors of our main extracts, together with those of a few other people. The material has been collected from many sources including the following:

BELL, E. T. *Men of mathematics.* Gollancz, London, 1937.

BOASE, F. *Modern English biography,* 6 vols. Netherton and Worth, Truro, 1892–1921.

MACFARLANE, A. *Lectures on ten British mathematicians of the nineteenth century.* Mathematical Monographs, No. 17, Wiley, New York, 1916.

MAY, K. O. *Bibliography and research manual of the history of mathematics.* University of Toronto Press, 1973.

POGGENDORF, J. C. *Biographisch-Literarisches Handworterbuch der exakten Naturwissenschaften,* Vols 1–7b, 1863–.

Dictionary of American biography, 20 vols, 1928-36, and supplementary vols.

Dictionary of national biography, 63 vols, 1885–1900, and supplementary vols.

Dictionary of scientific biography, 1970–.

Dictionary of South African biography, 2 vols, 1968–72.

Isis cumulative bibliography, 2 vols, 1971.

Meyers Lexikon, (7th edn.), 12 vols, 1924–30.

Neue deutsche Biographie, 1952–.

Various editions of: *Who was who, Who was who in America, American men and women of science, Encyclopaedia Britannica* and *Larousse.*

The biographies are arranged alphabetically and are preceded by a chronological list.

Euler, L.	1707–1783	Petersen, J. P. C.	1839–1910
Vandermonde, A.-T.	1735–1796	Hierholzer, C. F. B.	1840–1871
Lhuilier, S.-A.-J.	1750–1840	Tarry, G.	1843–1913
Poinsot, L.	1777–1859	Clifford, W. K.	1845–1879
Cauchy, A.-L.	1789–1857	Kempe, A. B.	1849–1922
Hamilton, W. R.	1805–1865	Heawood, P. J.	1861–1955
Kirkman, T. P.	1806–1895	Heffter, L. W. J.	1862–1962
De Morgan, A.	1806–1871	Veblen, O.	1880–1960
Listing, J. B.	1808–1882	Tietze, H. F. F.	1880–1964
Sylvester, J. J.	1814–1897	Birkhoff, G. D.	1884–1944
Cayley, A.	1821–1895	König, D.	1884–1944
Kirchhoff, G. R.	1824–1887	Pólya, G.	1887–
Frankland, E.	1825–1899	Kuratowski, K.	1896–
Tait, P. G.	1831–1901	Prüfer, E. P. H.	1896–1934
Guthrie, F.	1831–1899	Franklin, P.	1898–1965
Jordan, M.-E.-C.	1838–1922	Frink, O.	1901–
Crum Brown, A.	1838–1922	Whitney, H.	1907–

BIRKHOFF, George David   *b* 21 March 1884   Overisel, Mich.
                                d 12 November 1944   Cambridge, Mass.

After studying at the University of Chicago and at Harvard, Birkhoff held appointments at the University of Wisconsin and at Princeton before returning to Harvard in 1912 where he spent the rest of his career. He was generally regarded as the leading American mathematician of his time, and his interests included differential equations, difference equations, and dynamical systems; the contribution which he made to graph theory was a rather minor interest.

CAUCHY, Augustin-Louis   *b* 21 August 1789   Paris
                                d 22 May 1857   Sceaux

After training as a civil engineer Cauchy went, in 1810, to Cherbourg, where he worked on the design of the fortifications and harbour. His mathematical career began around this time, and one of his first papers was the memoir on polyhedra [5B].

Cauchy rapidly became the leading mathematician in France. His original work on groups of substitutions (permutations) and complex functions made him famous throughout the mathematical world, and he has more concepts and theorems named after him than any other mathematician. The influence of his lectures and books led to the introduction of proper standards of definition and proof in mathematical analysis.

CAYLEY, Arthur   *b* 16 August 1821   Richmond, Surrey
                        d 26 January 1895   Cambridge, England

After graduating from Cambridge in 1840 Cayley held a fellowship at Trinity College for a few years. He then entered the legal profession, and was called to the bar in 1849. By 1863, in spite of his legal duties, he had written nearly three hundred mathematical papers, and in that year he was appointed to the new Sadlerian chair of pure mathematics at Cambridge, a post which he held until his death.

In all, Cayley wrote nearly a thousand papers (he used to write out his discoveries and send them to be published without delay), but only one full-length book, a treatise on elliptic functions.

In 1881–2 Cayley lectured at Johns Hopkins University at the invitation of his close friend Sylvester, whom he had first encountered during his period as a lawyer.

CLIFFORD, William Kingdon   *b* 4 May 1845   Exeter
                                   d 3 March 1879   Madeira

Educated at King's College, London, and Trinity College, Cambridge, Clifford graduated in 1867. In 1870 he took part in an eclipse expedition to Italy and was shipwrecked in Sicily. In the following year he was appointed professor of applied mathematics and mechanics at University College, London. Clifford suffered a physical collapse in 1876 and the remainder of his life consisted of long visits to the Mediterranean area to recuperate. He left England in 1878 never to return.

Clifford's contemporaries believed that his early death deprived the world of one of the best mathematicians of his time.

CRUM BROWN, Alexander   *b* 26 March 1838   Edinburgh
                                       d 28 October 1922   Edinburgh

Crum Brown graduated at Edinburgh University in 1858, received his M.D. in 1861 and a D.Sc. from London in the following year. After spending several months in Germany, he returned to Edinburgh as a lecturer in chemistry in 1863. He was promoted to professor in 1869, and retired in 1908. Crum Brown was a colleague and brother-in-law of Tait.

DE MORGAN, Augustus   *b* 27 June 1806   Madura, India
                                       d 18 March 1871   London

De Morgan studied at Trinity College, Cambridge, and graduated in 1827. He was known as an eccentric and renowned for his strongly held beliefs. In 1828 he was appointed professor of mathematics at what is now University College, London, but resigned on a matter of principle in 1831. He was reappointed in 1836, but resigned again in 1866.

The most important researches of De Morgan were in the field of logic, but he also took an interest in the history of mathematics and formed a large collection of arithmetical books. He wrote many articles for the *Penny cyclopaedia* (1833–43) and contributed regularly to several periodicals. He was also a strong advocate of decimal currency.

One of De Morgan's best-known books, the *Budget of paradoxes*, was edited by his widow and published posthumously in 1872.

EULER, Leonhard   *b* 15 April 1707   Basel
                                       d 18 September 1783 St. Petersburg

Euler's father was a Calvinist pastor who wished his son to follow a career in the church, so Leonhard entered the University of Basel to study theology and Hebrew. Eventually his father was persuaded that Leonhard was destined to become a mathematician, and his theological training was abandoned.

In 1727 Euler was called to the Academy at St. Petersburg officially as an associate of the medical section, but in effect he soon became a member of the mathematical section. Six years later he succeeded Daniel Bernoulli as the leading mathematician in the Academy.

In 1740 Euler moved to the Berlin Academy, but his years spent there were not entirely happy and he was glad to be able to return to St. Petersburg in 1766.

Several of the mathematicians mentioned in our book were prolific authors, but Euler surpassed them all. He lost the sight of one eye during his first period in Russia, and became totally blind soon after returning there, but this seems to have increased his output. He performed extensive calculations in his head and dictated the results to amanuenses, including his son Johann Albrecht.

Euler died suddenly in his seventy-seventh year; he had remained active to the end and was still performing calculations on the day of his death.

FRANKLAND, Edward   *b* 18 January 1825   Catterall, Lancs.
                                       d 9 August 1899   Golaa, Norway

Frankland was apprenticed to a Lancashire druggist who taught him little, and then he secured employment at the Museum of Economic Geology in London. He subsequently held various appointments both in Britain and in continental Europe before returning to London in 1863. His duties included making the official monthly analyses of the water supplies of London, and he served on the second Royal Commission on River Pollution. He was knighted in 1897.

FRANKLIN, Philip   *b* 5 October 1898   New York City
                                          d 27 January 1965 Boston, Mass.

After receiving his Ph.D. from Princeton in 1921, Franklin was an instructor in mathematics at Princeton and Harvard before moving to the Massachusetts Institute of Technology in 1924. He became a full professor there in 1937. Besides his work on map colouring, he published books on calculus, differential equations, complex variable, and Fourier methods.

FRINK, Orrin   *b* 31 May 1901   Brooklyn, New York

After receiving his Ph.D. from Columbia University, Frink spent the academic year 1925–6 as an instructor at Princeton University where he attended a seminar conducted by Veblen. He was given the assignment of applying Veblen's methods to the four-colour problem, and in particular to simplifying Brahana's proof of Petersen's theorem. In the event, Frink realised that a much simpler proof was possible, and the outcome was his paper [10C].

In 1928 Frink moved to Pennsylvania State University where he remained until his retirement. He has published papers on a variety of topics including lattice theory and topology.

GUTHRIE, Francis   *b*   1831   London
                                          d 19 October 1899   Cape Town

After obtaining a B.A. degree from University College, London, Guthrie went on to obtain a law degree and to practice as a barrister. In 1861 he was appointed professor of mathematics in the newly established Graaff-Reinet College, Cape Colony, and moved to Cape Town in 1875; he obtained a chair in mathematics at the South African College in the following year.

Guthrie's contribution to graph theory was, in some respects, minimal; he conjectured one result (the four-colour problem), but made no direct contribution to the theory.

Guthrie published a few mathematical papers, but he was also interested in botany. In particular he is remembered for his work on the genus *Erica*, in which he is commemorated by *Erica Guthriei*.

HAMILTON, William Rowan   *b* midnight 3–4 August 1805   Dublin
                                          d   2 September 1865   Dunsink

Hamilton was a child prodigy, and by the age of 14 had mastered several languages, modern, classical, and oriental. He also showed great ability in mathematics, and in 1822 noticed an important error in Laplace's *Mécanique céleste*. As a result of this he came to the attention of the Astronomer Royal of Ireland. Hamilton entered Trinity College, Dublin, at the age of 18, and was an outstanding student in both classics and science. Before he had taken his degree he was appointed Astronomer Royal of Ireland and took up his duties in 1827. He remained at Dunsink Observatory until his death.

Hamilton was not a success as a practical astronomer, but his theoretical work in geometrical optics, dynamics, and algebra was brilliant. He had an obsession for his work on quaternions, and although this particular topic proved less useful than he hoped, his discovery of non-commutative systems has revolutionized algebra.

The later career of Hamilton was marred by his addiction to alcohol, and he died from an acute attack of gout.

HEAWOOD, Percy John   *b*   8 September 1861   Newport, Salop
                                      d 24 January 1955   Durham

Heawood studied at Oxford and remained there until 1887 when he was appointed a lecturer at the Durham Colleges, later to become Durham University. He remained in Durham for the rest of his long life.

He has earned a place in history on two counts. The first of these was his work on map-colouring which began with his paper [6D/7A] and extended over a period of nearly sixty years. The other was his work as secretary of the Durham Castle restoration fund. In 1928 the castle's foundations were found to be insecure, and long after others had given up hope of ever raising sufficient funds to prevent it from sliding down the hill on which it stands, Heawood persevered. His successful efforts were rewarded by honours from both his University and his Country.

HEFFTER, Lothar Wilhelm Julius   *b* 11 June 1862   Köslin
                                        d   1 January 1962   Freiburg im
                                                                Breisgau

After studying mathematics and physics at Heidelberg and Berlin, Heffter held various appointments at Giessen, Bonn, Aachen, and Kiel. He eventually settled down at Freiburg in 1911, where he was to remain for over half a century until his death. Although he became an emeritus professor in 1931, he continued lecturing regularly until 1936 and began lecturing again during the Second World War because of the shortage of teachers. His teaching duties finally came to an end with the devastation of Freiburg by bombing in 1944.

Heffter wrote a short autobiography when he was 75 (*Mein Lebensweg und meine Mathematische Arbeit*, Leipzig, 1937) and at the age of 90 a much longer one (*Beglückte Ruchschau auf 9 Jahrzehnte*, Freiburg, 1952).

HIERHOLZER, Carl Fridolin Bernhard   *b*   2 October 1840   Freiburg im
                                                                   Breisgau
                                         d 13 September 1871 Karlsruhe

The short life of Hierholzer included studies in Karlsruhe, Berlin, and Heidelberg, where he was awarded his doctorate in 1865. In 1870 he became assistant to Lüroth, an algebraic geometer, in Karlsruhe. He has three papers in the early volumes of *Mathematische Annalen*, the last one being his posthumous paper on Eulerian paths [1B].

JORDAN, Marie-Ennemond-Camille   *b*   5 January 1838   Lyons
                                              d 20 January 1922   Paris

Camille Jordan followed the usual career of French mathematicians of the nineteenth century. He trained as an engineer at the École Polytechnique and did mathematics in his spare time. Subsequently (1873–1912) he taught at both the École Polytechnique and the Collège de France.

He published papers in many areas of mathematics and is especially remembered for his *Cours d'analyse de l'École Polytechnique* which reached a third edition and had a great influence on the development of analysis. Equally important was his work in algebra, and his *Traité des substitutions et des équations algébriques* was the standard work in group theory for some thirty years. His one contribution to graph theory [3B] was concerned with the symmetry groups of graphs and trees.

KEMPE, Alfred Bray   *b*   6 July 1849   London
                                       d 21 April 1922   London

Kempe was a barrister by profession and became a leading authority on Ecclesiastical Law. But he had studied mathematics at Cambridge, and through-out his life he continued to devote some of his time and energy to it. His first researches were in kinematics, in particular the construction of linkages for drawing various curves. The result was a minor classic entitled *How to draw a straight line*. Another of Kempe's works was *A memoir on the theory of mathematical form*, published in 1886. Although his ideas on this subject have been superseded by later work on mathematical logic, his use of a graphical notation in this context is worthy of note. Strangely, the one work which assures Kempe a permanent place in the history of mathematics was his paper [6B] containing a fallacious 'proof'!

For many years Kempe held the post of treasurer of the Royal Society. He was knighted in 1912.

KIRCHHOFF, Gustav Robert   *b* 12 March 1824   Königsberg
                                      d 17 October 1887 Berlin

At the age of 21, whilst studying at the University of Königsberg, Kirchhoff formulated his famous laws on the flow of electric currents. He then spent three years as a Privatdozent at the University of Berlin before accepting the post of professor of physics at Breslau, to which he moved in 1850. Four years later he moved to Heidelberg where, as a result of work done there, he propounded a fundamental law of electromagnetic radiation.

For a number of years Kirchhoff suffered from a disability which forced him to use crutches or a wheelchair. Eventually his failing health hindered his experimental work, and, in 1875, he was appointed professor of theoretical physics at Berlin.

KIRKMAN, Thomas Penyngton   *b* 31 March 1806   Bolton
                                     d   3 February 1895 Croft, Lancashire

After studying at Dublin University, Kirkman entered the Church of England. After two brief curacies he became rector of the parish of Croft with Southworth in Lancashire, where he was to remain for fifty-two years.

Besides some theological pamphlets, Kirkman published a remarkable number of mathematical papers. His early work was on combinatorial puzzles, and his name is associated with the famous problem of the *fifteen schoolgirls*. In 1853 he began to publish papers on what he always called 'polyedra', and he retained his interest in that topic for the rest of his life. His discovery of 'Hamiltonian' circuits [2B] is just one of the many significant ideas for which he has not received due credit.

KÖNIG, Dénes   *b* 21 September 1884   Budapest
                                              d 19 October 1944   Budapest

Dénes König was the son of the mathematician Julius König, whose posthumously published book *Neue Grundlagen der Logik, Arithmetik und Mengenlehre* was edited by Dénes and published in 1914.

König studied at Budapest and Göttingen, receiving his doctorate from Budapest in 1907. The whole of his career was also spent at Budapest, initially as an assistant and finally as a professor. His life came to a tragic end in 1944 when he committed suicide.

Most of König's work was in the field of combinatorics and he wrote the first comprehensive treatise on graph theory, *Theorie der endlichen und unendlichen Graphen*, published in 1936.

KURATOWSKI, Kazimierz   *b* 2 February 1896   Warsaw

Kuratowski studied in Glasgow, where he read engineering, and in Warsaw. From 1921 to 1934 he worked in Lvov, and then returned to Warsaw. His main interest was analytic topology, and the paper [8C] which earned him a permanent place in graph theory was written from the standpoint of that subject.

LHUILIER, Simon-Antoine-Jean   *b* 24 April 1750   Geneva
                                              d 28 March 1840 Geneva

Lhuilier began his studies of mathematics and physics at Calvin Academy. In 1775 he entered a competition which was being held by the commission in charge of preparing textbooks for use in Polish schools. His proposal for a mathematics text won the competition, and as a result Lhuilier was asked to become a tutor in the household of a Polish royal family. Despite the social obligations of this post, he wrote the proposed text and several memoirs.

After leaving Poland in 1789, Lhuilier spent most of the next few years in Tübingen and then held the chair of mathematics at the Geneva Academy from 1795 until his retirement in 1823.

LISTING, Johann Benedict   *b* 25 July 1808   Frankfurt am Main
                                              d 24 December 1882   Göttingen

There seem to be few accounts of the life of Listing, but what little there is suggests that he was an original thinker who deserves to be much better known. He was of Czech descent, began his studies in mathematics at Göttingen in 1829, and graduated in 1834. He spent most of the next three years in Italy and Sicily, where he helped a colleague perform scientific experiments. After two years in Hannover, Listing returned to Göttingen in 1839 as professor of mathematical physics and optics.

In addition to the two mathematical classics, the *Topologie* [1C] and the *Census*, Listing wrote pioneering work on physiological optics and was awarded an honorary doctorate of medicine by the University of Tübingen in 1877.

PETERSEN, Julius Peter Christian   *b* 16 June 1839   Sorø
                                              d 5 August 1910   Copenhagen

Petersen was an important figure in Danish mathematics. He obtained his Ph.D. in 1871, and taught in a polytechnic and a military academy, later becoming a professor at the University of Copenhagen. He was a member of the board of inspectors of senior schools (1887–1900) and was nationally known as the author of a series of school and undergraduate textbooks, one of which, *Methods and theories for the solution of problems of geometrical constructions*, achieved international recognition, being translated into five languages. He was, however, interested only in able students and his books for elementary schools proved to be too advanced.

Petersen's research interests were wide and he wrote on algebra, number theory, analysis, mechanics, and geometry. His most important work was in geometry, and his paper on regular graphs [10A] was perhaps the most important of all.

POINSOT, Louis   *b* 3 January 1777 Paris
                 d 5 December 1859 Paris

The École Polytechnique was established in 1794, and Poinsot was a member of its first graduation class, in 1797. He qualified as a civil engineer and was soon appointed professor at the Lycée Bonaparte. From 1816 to 1825 he was the first professor of analysis and mechanics at the École Polytechnique, and he later became entrance examiner. His subsequent appointments seem to have been in administration rather than in teaching.

One of his earliest works was *Éléments statique* (1803) and, in addition to his work on geometry (1809) mentioned in Chapter 1, his researches included the history of numbers.

PÓLYA, George   *b* 13 December 1887 Budapest

The son of a statistician, Pólya studied at Budapest, Göttingen, Vienna, and Paris, receiving his Ph.D. from Budapest in 1912. From 1914 to 1940 he taught in Zürich, initially as Privatdozent and later as professor.

In 1940 Pólya emigrated to the United States where he was visiting professor at Brown University until 1942. Then, after a brief period at Smith College, Massachusetts, he moved to Stanford where he remained, becoming emeritus professor in 1953.

Pólya has worked in several areas of mathematics and, in addition, has established a fine reputation in mathematical education.

PRÜFER, Ernst Paul Heinz   *b* 10 November 1896 Wilhelmshaven
                d 7 April 1934 Münster

After receiving his doctorate from Berlin in 1921, Prüfer was a Privatdozent at Jena and then at Münster. He became a professor at Münster in 1930 and died four years later.

Besides his paper [3D] on permutations, Prüfer published work on Abelian groups, algebraic numbers, Sturm–Liouville theory, and knot theory. From his papers a book on projective geometry was compiled by G. Fleddermann and G. Köthe, and published privately.

SYLVESTER, James Joseph   *b* 3 September 1814 London
                d 15 March 1897 London

Sylvester studied at St. John's College, Cambridge. He held a brief appointment at the University of Virginia in 1841, and after returning to England was called to the bar. During this period of his life he became friendly with Cayley, and they began their famous work on the algebraic theory of invariants.

In 1855 he became professor of mathematics at the Royal Military Academy, Woolwich, but War Office regulations retired him on half pay when he was 55. In 1876 Sylvester returned to the United States to become professor at Johns Hopkins University. There he founded the *American Journal of Mathematics*, the first volume of which contained his thoughts on the relationship between chemistry and algebra.

At the age of 68, Sylvester returned to his native land to become Savilian professor of geometry at Oxford.

TAIT, Peter Guthrie   *b* 28 April 1831   Dalkeith
                                             d 4 July 1901   Edinburgh

After attending Edinburgh Academy, Tait spent one session at Edinburgh before moving to Peterhouse, Cambridge, where he graduated in 1852. He remained at Peterhouse as a fellow until 1854, when he was appointed professor of mathematics at Queen's College, Belfast. But Tait's interests were inclined more towards the real world than to pure mathematics, and in 1860 he became professor of natural philosophy at Edinburgh; he was to remain there until his retirement in 1901, shortly before his death.

Tait was an accomplished teacher and prolific author whose writings included a substantial number of articles for the *Encyclopaedia Britannica* and *Chambers' encyclopaedia*.

TARRY, Gaston   *b* 27 September 1843   Villefranche, Aveyron
                                  d 21 June 1913   Le Havre

Tarry studied at the École Polytechnique in 1857, and later became a finance inspector. Much of his life was spent in Algeria. His best-known papers are on the maze problem [1D], and on Euler's problem of the thirty-six officers; in modern terminology, the latter paper proved the non-existence of a pair of orthogonal 6×6 Latin squares.

TIETZE, Heinrich Franz Friedrich   *b* 31 August 1880   Schleinz
                                            d 17 February 1964   Munich

Tietze studied at Vienna, Munich, and Göttingen, and received his doctorate from Vienna in 1904. He became a Privatdozent in Vienna in 1908, and was later appointed professor at Brunn (1910–9), Erlangen (1919–25) and then Munich. He became emeritus professor in 1950, and continued to live in Munich.

Besides his paper [7C] on map colourings, his works include a well-known book called *Famous problems of mathematics*, which contains some fine colour plates illustrating map colourings on certain surfaces.

VANDERMONDE, Alexandre-Theophile   *b* 28 February 1735 Paris
                                            d 1 January 1796   Paris

Music was Vandermonde's first love, and he turned to mathematics relatively late in life. He entered the Académie des Sciences, Paris, in 1771, and his entire contribution to mathematics consists of four memoirs published in the *Histoires* of the Academy in the years 1771–2. The most important of these four is a treatment of the symmetric functions of roots of equations. The others are concerned with factorials, determinants, and the knight's tour [2A]. (It is worth remarking that the paper on determinants makes no mention of the so-called Vandermonde determinant.)

VEBLEN, Oswald   *b* 24 June 1880   Decorah, Iowa
                                          d 10 August 1960   Brooklin, Maine

Veblen's distinguished career began with studies in Iowa, Harvard and the University of Chicago. After teaching for two years in Chicago, he moved to Princeton University (1905–32), and later to the Institute of Advanced Study, Princeton, where he played an important role in organizing the school of mathematics.

His early work on the foundations of geometry led him to write two volumes entitled *Projective geometry* (Vol. 1 with J. W. Young). His *Analysis situs* (see [8B]) was his most important book and the standard work on topology for many years. Another important book, which he wrote with his student, Henry Whitehead, was *Foundations of differential geometry.*

Veblen's interest in geometry led him to try to construct a theory to unify gravitation and electromagnetism. An offshoot of this work was a new treatment of spinors included in his book *Projektive Relativitätstheorie.*

WHITNEY, Hassler   *b* 23 March 1907   New York City

After receiving the degrees of Ph.B. and Mus.B from Yale, Whitney moved to Harvard where he received his Ph.D. in 1932. Apart from a year at Princeton, he remained at Harvard until 1952, becoming full professor in 1946. He then moved to the Institute of Advanced Study, Princeton, where he has remained ever since.

Whitney was a grandson of Simon Newcomb, a contemporary of Sylvester at Johns Hopkins University. He produced more than a dozen papers on graph theory, a subject in which he became interested during a visit to Germany, and has also worked on manifolds, integration theory, and analytic varieties.

APPENDIX 3

Bibliography: 1736–1936

THIS appendix contains a chronological list of papers on graph theory covering the years 1736–1936. Journals relating to the activities of learned societies and published in the eighteenth and nineteenth centuries often bear two dates: one relating to the year in which the meetings were held and the other a publisher's date. As these dates differ by up to as much as five years it has been difficult to decide precisely where to place some papers in the list, but we have generally preferred the earlier dates.

We have had to be selective at times. For example, there seemed little sense in including every reference to the Königsberg bridge problem, or to the knight's tour problem, merely because such problems can be given graph-theoretical form.

Where the original article was untitled, we have supplied a title but inserted it in square brackets; in a few cases we have amplified a title and the addition is also bracketed.

There seems to be no standard list of journal abbreviations which covers all the titles in our bibliography. Where possible we have followed the list appearing in the *Isis cumulative bibliography*, Vol. 1, Mansell (in conjunction with the History of Science Society), London, 1971.

Appropriate references are included to the following collected works:

BIRKHOFF, G. D. *Collected mathematical papers*, 3 vols. American Mathematical Society, New York, 1950.

CAUCHY, A.-L. *Oeuvres complètes d'Augustin Cauchy*, 27 vols. Gauthier-Villars, Paris, 1882–1970.

CAYLEY, A. *Mathematical papers*, 13 vols and index vol. Cambridge University Press, 1889–97.

CLIFFORD, W. K. *Mathematical papers*. Macmillan, London, 1882.

EULER, L. *Opera omnia, series prima, opera mathematica*, 29 vols. Sub auspiciis societatis scientiarum naturalium Helveticae, 1911–56.

HAMILTON, W. R. *Mathematical papers of Sir William Rowan Hamilton*, Vol. 3. Cambridge University Press, 1967.

JORDAN, C. *Oeuvres de Camille Jordan*, 4 vols. Gauthier-Villars, Paris, 1961–4.

KIRCHHOFF, G. R. *Gesammelte Abhandlungen von G. Kirchhoff*. Barth, Leipzig, 1882.

MAXWELL, J. C. *Scientific papers of James Clerk Maxwell*, 2 vols. Cambridge University Press, 1890.

MÖBIUS, A. F. *August Ferdinand Möbius Gesammelte Werke*, 4 vols. Hirzel, Stuttgart, 1885–7.

POINCARÉ, H. *Oeuvres de Henri Poincaré*, 11 vols. Gauthier-Villars, Paris, 1916–56.

SYLVESTER, J. J. *Collected mathematical papers of James Joseph Sylvester*, 4 vols. Cambridge University Press, 1904–12.

TAIT, P. G. *Scientific papers*, 2 vols. Cambridge University Press, 1898–1900.

1736

EULER, L. Solutio problematis ad geometriam situs pertinentis. *Comm. Acad. Sci. Imp. Petropol.* **8** (1736), 128–140 = *Opera Omnia* (1), Vol. 7, 1–10. For translations, see [1A] (English), Coupy 1851 (French), and Speiser 1925 (German).

1752

EULER, L. Elementa doctrinae solidorum. *Novi Comm. Acad. Sci. Imp. Petropol.* **4** (1752–3), 109–140 = *Opera Omnia* (1), Vol. 26, 72–93.

EULER, L. Demonstratio nonnullarum insignium proprietatum quibus solida hedris planis inclusa sunt praedita. *Novi Comm. Acad. Sci. Imp. Petropol.* **4** (1752–3), 140–160 = *Opera Omnia* (1), Vol. 26, 94–108.

1759

EULER, L. Solution d'une question curieuse qui ne paroit soumise à aucune analyse. *Mém. Acad. Sci. Berlin* **15** (1759), 310–337 = *Opera Omnia* (1), Vol. 7, 26–56.

1771

VANDERMONDE, A.-T. Remarques sur les problèmes de situation. *Hist. Acad. Sci. (Paris)* (1771), 566–574. [2A]

1810

POINSOT, L. Sur les polygones et les polyèdres. *J. École Polytech.* **4** (Cah. 10) (1810), 16–48.

1811

L'HUILIER, S. Démonstration immédiate d'un théorème fondamental d'Euler sur les polyhèdres, et exceptions dont ce théorème est susceptible. *Mém. Acad. Imp. Sci. St₄ Pétersb.* **4** (1811), 271–301.

1812

LHUILIER, S. (abridged by J. D. Gergonne). Mémoire sur la polyédrométrie; contenant une démonstration directe du théorème d'Euler sur les polyèdres, et un examen des diverses exceptions auxquelles ce théorème est assujetti. *Ann. de Math.* **3** (1812–3), 169–189. [5C]

1813

CAUCHY, A.-L. Recherches sur les polyèdres—premier mémoire. *J. École Polytech.* **9** (Cah. 16) (1813), 68–86 = *Oeuvres* (2), Vol. 1, 7–25. [5B]

1827

GRUNERT, J. A. Einfacher Beweis der von Cauchy und Euler gefundenen Sätze von Figurennetzen und Polyëdern. *J. Reine Angew. Math.* **2** (1827), 367.

1844

CLAUSEN, T. [Second postscript to:] De linearum tertii ordinis proprietatibus. *Astron. Nachr.* **21** (1844), col. 209–216.

1847

KIRCHHOFF, G. R. Über die Auflösung der Gleichungen, auf welche man bei der Untersuchung der linearen Vertheilung galvanischer Ströme geführt wird. *Ann. Phys. Chem.* **72** (1847), 497–508 = *Ges. Abh.*, 22–33. [8A]

LISTING, J. B. Vorstudien zur Topologie. *Göttinger Studien (Abt. 1) Math. Naturwiss. Abh.* **1** (1847), 811–875 = Separately as book, Göttingen, 1848. [1C]

1849

TERQUEM, O. Sur les polygones et les polyèdres étoilés, polygones funiculaires; d'après M. Poinsot. *Nouv. Ann. Math.* **8** (1849), 68–74.

TERQUEM, O. Polyèdres réguliers ordinaires et polyèdres réguliers étoilés; d'après M. Poinsot. *Nouv. Ann. Math.* **8** (1849), 132–139.

1851

COUPY, É. Solution d'un problème appartenant a la géométrie de situation, par Euler. *Nouv. Ann. Math.* **10** (1851), 106–119.

1855

KIRKMAN, T. P. On the representation and enumeration of polyedra. *Mem. Lit. Phil. Soc. Manchester* (2) **12** (1855), 47–70.

1856

HAMILTON, W. R. Memorandum respecting a new system of roots of unity. *Phil. Mag.* (4) **12** (1856), 446 = *Math. Papers*, Vol. 3, 610.

HAMILTON, W. R. [Account of the icosian calculus]. *Proc. Roy. Irish. Acad.* **6** (1853–7), 415–416 = *Math. Papers*, Vol. 3, 609.

KIRKMAN, T. P. On the enumeration of x-edra having triedral summits and an $(x-1)$-gonal base. *Phil. Trans. Roy. Soc. London* **146** (1856), 399–411.

KIRKMAN, T. P. On the representation of polyedra. *Phil. Trans. Roy. Soc. London* **146** (1856), 413–418. [2B]

1857

CAYLEY, A. On the theory of the analytical forms called trees. *Phil. Mag.* (4) **13** (1857), 172–176 = *Math. Papers*, Vol. 3, 242–246. [3A]

KIRKMAN, T. P. On autopolar polyedra. *Phil. Trans. Roy. Soc. London* **147** (1857), 183–215.

1858

KIRKMAN, T. P. On the partitions of the R-pyramid, being the first class of R-gonous X-edra. *Phil. Trans. Roy. Soc. London* **148** (1858), 145–161.

POINSOT, L. Note sur la théorie des polyèdres. *Compt. Rend. Acad. Sci. Paris* **46** (1858), 65–79.

1859

CAYLEY, A. On the theory of the analytical forms called trees—part II. *Phil. Mag.* (4) **18** (1859), 374–378 = *Math. Papers*, Vol. 4, 112–115.

1861

CAYLEY, A. On the partitions of a close. *Phil. Mag.* (4) **21** (1861), 424–428 = *Math. Papers*, Vol. 5, 62–65.

LISTING, J. B. Der Census räumlicher Complexe oder Verallgemeinerung des Euler'schen Satzes von den Polyëdern. *Abh. K. Ges. Wiss. Göttingen Math. Cl.* **10** (1861–2), 97–182 = Separately as book, Dieterich'sche Verlagshandlung, Göttingen, 1862.

LISTING, J. B. Der Census räumlicher Complexe. *Nachr. K. Ges. Wiss. Göttingen* (1861), 352–358.

1862

CAYLEY, A. On the △ faced polyacrons, in reference to the problem of the enumeration of polyhedra. *Mem. Lit. Phil. Soc. Manchester* **1** (1862), 248–256 = *Math. Papers*, Vol. 5, 38–43.

HERSCHEL, A. S. Sir Wm. Hamilton's icosian game. *Quart. J. Pure Appl. Math.* **5** (1862), 305.

1864

MAXWELL, J. C. On reciprocal figures and diagrams of forces. *Phil. Mag.* (4) **27** (1864), 250–261 = *Sci. Papers*, Vol. 1, 514–525.

1867

Бугаев, Н. В. Теорема Эйлера о многогранникахъ. Свойства плоской геометрической сети. *Mat. Sb.* **2** (1867), 87–92.

Listing, J. B. Über einige Anwendungen des Census-Theorems. *Nachr. K. Ges. Wiss. Göttingen* (1867), 430–447 = *Arch. Math. Phys.* **48** (1868), 186–198.

1869

Jordan, C. Sur les assemblages de lignes. *J. Reine Angew. Math.* **70** (1869), 185–190 = *Oeuvres*, Vol. 4, 303–308. [3B]

1870

Maxwell, J. C. On reciprocal figures, frames, and diagrams of forces. *Trans. Roy. Soc. Edinb.* **26** (1869–72), 1–40 = *Sci. Papers*, Vol. 2, 161–207.

1873

Cayley, A. On Listing's theorem. *Mess. Math.* (New Ser.) **2** (1873), 81–89 = *Math. Papers*, Vol. 8, 540–547.

Hierholzer, C. Über die Möglichkeit, einen Linienzug ohne Wiederholung und ohne Unterbrechnung zu umfahren. *Math. Ann.* **6** (1873), 30–32. [1B]

Wiener, C. Über eine Aufgabe aus der Geometria situs. *Math. Ann.* **6** (1873), 29–30.

1874

Cayley, A. On the mathematical theory of isomers. *Phil. Mag.* (4) **47** (1874), 444–446 = *Math. Papers*, Vol. 9, 202–204. [4B]

Sylvester, J. J. On recent discoveries in mechanical conversion of motion. *Proc. Roy. Inst. Gt. Brit.* **7** (1873–5), 179–198 = *Math. Papers*, Vol. 3, 7–25.

1875

Cayley, A. Über die analytischen Figuren, welche in der Mathematik Bäume genannt werden und ihre Anwendung auf die Theorie chemischer Verbindungen. *Ber. Deut. Chem. Ges.* **8** (1875), 1056–1059.

Cayley, A. On the analytical forms called trees, with application to the theory of chemical combinations. *Rep. Brit. Assoc. Advance. Sci.* **45** (1875), 257–305 = *Math. Papers*, Vol. 9, 427–460.

Schiff, H. Zur Statistik chemischer Verbindungen. *Ber. Deut. Chem. Ges.* **8** (1875), 1542–1547.

1876

Saalschütz, L. [Untitled]. *Schr. Phys.-Ökon. Ges. Königsberg Prussia* **16** (1876), 23–24.

1877

Cayley, A. [Solution to question 5208]. *Math. Quest. Solut. Educ. Times* **27** (1877), 81–83 = *Math. Papers*, Vol. 10, 598–600.

Cayley, A. On the number of the univalent radicals C_nH_{2n+1}. *Phil. Mag.* (5) **3** (1877), 34–35 = *Math. Papers*, Vol. 9, 544–545.

Sylvester, J. J. Question 5208. *Math. Quest. Solut. Educ. Times* **27** (1877), 81.

1878

Cayley, A. The theory of groups: graphical representation. *Amer. J. Math.* **1** (1878), 174–176 = *Math. Papers*, Vol. 10, 403–405.

CAYLEY, A. On the theory of groups. *Proc. London Math. Soc.* **9** (1877–8), 126–133 = *Math. Papers*, Vol. 10, 324–330.

CLIFFORD, W. K. Extract of a letter to Mr. Sylvester.... *Amer. J. Math.* **1** (1878), 126–128 = *Math. Papers*, 255–257.

LISTING, J. B. [Der sogenannte Census–Satz]. *Amtl. Ber. Versamm. Deut. Natur. Ärzte* (1878), 224–227.

SYLVESTER, J. J. On an application of the new atomic theory to the graphical representation of the invariants and covariants of binary quantics,—with three appendices. *Amer. J. Math.* **1** (1878), 64–125 = *Math. Papers*, Vol. 3, 148–206.

SYLVESTER, J. J. Chemistry and algebra. *Nature* **17** (1877–8), 284 = *Math. Papers*, Vol. 3, 103–104. [4C]

1879

CAYLEY, A. On the colouring of maps. *Proc. Roy. Geog. Soc.* (New Ser.) **1** (1879), 259–261 = *Math. Papers*, Vol. 11, 7–8. [6A]

CLIFFORD, W. K. Binary forms of alternate variables. *Proc. London Math. Soc.* **10** (1878–9), 214–221 = *Math. Papers*, 277–286.

KEMPE, A. B. On the geographical problem of the four colours. *Amer. J. Math.* **2** (1879), 193–200. [6B]

KEMPE, A. B. [Untitled abstract]. *Proc. London Math. Soc.* **10** (1878–9), 229–231.

SPOTTISWOODE, W. On Clifford's graphs. *Proc. London Math. Soc.* **10** (1878–9), 204–214.

STORY, W. E. Note on the preceding paper [by Kempe]. *Amer. J. Math.* **2** (1879), 201–204.

SYLVESTER, J. J. On the mathematical question, what is a tree? *Math. Quest. Solut. Educ. Times* **30** (1879), 52.

SYLVESTER, J. J. Question 573[8]. *Math. Quest. Solut. Educ. Times* **30** (1879), 81. [Incorrectly numbered 573; 5738 in *Educ. Times* itself].

TEBAY, S. [Solution to question 5738]. *Math. Quest. Solut. Educ. Times* **30** (1879), 81–85.

1880

DE POLIGNAC, C. Formules et considérations diverses se rapportant à la théorie des ramifications. *Bull. Soc. Math. France* **8** (1879–80), 120–124.

DE POLIGNAC, C. Formules et considérations diverses se rapportant à la théorie des ramifications. *Bull. Soc. Math. France* **9** (1880–81), 30–42.

GUTHRIE, F. Note on the colouring of maps. *Proc. Roy. Soc. Edinb.* **10** (1880), 727–728.

KEMPE, A. B. How to colour a map with four colours. *Nature* **21** (1879–80), 399–400.

TAIT, P. G. On the colouring of maps. *Proc. Roy. Soc. Edinb.* **10** (1878–80), 501–503.

TAIT, P. G. Remarks on the previous communication [by Guthrie]. *Proc. Roy. Soc. Edinb.* **10** (1878–80), 729. [6C]

TAIT, P. G. Note on a theorem in the geometry of position. *Trans. Roy. Soc. Edinb.* **29** (1880), 657–660 = *Sci. Papers*, Vol. 1, 408–411.

1881

CAYLEY, A. On the analytical forms called trees. *Amer. J. Math.* **4** (1881), 266–268 = *Math. Papers*, Vol. 11, 365–367. [3C]

CLIFFORD, W. K. *Mathematical fragments being facsimiles of his unfinished papers relating to the theory of graphs.* Macmillan, London, 1881.

KIRKMAN, T. P. Question 6610, solution by the proposer. *Math. Quest. Solut. Educ. Times* **35** (1881), 112–116.

LEMOINE, É. Quelques questions de géométrie de position sur les figures qui peuvent se tracer d'un seul trait. *Compt. Rend. Ass. Franç. Avance. Sci.* **10** (1881), 175–180.

1882

CAYLEY, A. On the geometrical forms called trees. *Johns Hopkins Univ. Circ.* **1** (1879–82), 202.

LUCAS, É. *Récréations mathématiques*, 4 vols. Gauthier-Villars, Paris, 1882–94.

SYLVESTER, J. J. [On the geometrical forms called trees]. *Johns Hopkins Univ. Circ.* **1** (1879–82), 202–203 = *Math. Papers*, Vol. 3, 640–641.

1883

KIRKMAN, T. P. On the enumeration and construction of polyedra whose summits are all triedral, and which have neither triangle nor quadrilateral. *Proc. Lit. Phil. Soc. Liverpool* **37** (1882–3), 49–66.

LUCAS, É. Le problème géographique des quatre couleurs. *Rev. Sci.* (3) **6** (1883), 12–17.

1884

TAIT, P. G. Listing's *Topologie. Phil. Mag.* (5) **17** (1884), 30–46 = *Sci. Papers*, Vol. 2, 85–98.

1885

BALTZER, R. Eine Erinnerung an Möbius und seinen Freund Weiske. *Ber. K. Sächs. Ges. Wiss. Leipzig Math.-Phys. Cl.* **37** (1885), 1–6.

1886

KEMPE, A. B. A memoir on the theory of mathematical form. *Phil. Trans. Roy. Soc. London* **177** (1886), 1–70.

TARRY, G. Nombre de manières distinctes de parcourir en une seule course toutes les allées d'un labyrinthe rentrant, en ne passant qu'une seule fois par chacune des allées. *Compt. Rend. Ass. Franç. Avance. Sci.* **15** Pt. 2 (1886), 49–53.

1889

CAYLEY, A. On the theory of groups. *Amer. J. Math.* **11** (1889), 139–157 = *Math. Papers*, Vol. 12, 639–656.

CAYLEY, A. A theorem on trees. *Quart. J. Pure Appl. Math.* **23** (1889), 376–378 = *Math. Papers*, Vol. 13, 26–28.

1890

BRUNEL, G. [Coloration d'une carte géographique]. *Mém. Soc. Sci. Phys. Nat. Bordeaux* (3) **5** (1890), *Extr. Proc.-Verb.* lxxxix.

HEAWOOD, P. J. Map-colour theorem. *Quart. J. Pure. Appl. Math.* **24** (1890), 332–338. [6D/7A]

1891

BRUNEL, G. Sur les configurations régulières tracées sur une surface quelconque. *Mém. Soc. Sci. Phys. Nat. Bordeaux* (4) **2** (1891), *Extr. Proc.-Verb.* xxii–xxiv.

BRUNEL, G. Sur la relation qui existe entre le problème de la détermination de la distribution électrique dans un système de conducteurs linéaires et la question du saut du cavalier sur l'échiquier. *Mém. Soc. Sci. Phys. Nat. Bordeaux* (4) **2** (1891), *Extr. Proc.-Verb.* xxxiv–xxxvi.

HEFFTER, L. Über das Problem der Nachbargebiete. *Math. Ann.* **38** (1891), 477–508. [7B]

MACMAHON, P. A. Yoke-chains and multipartite compositions in connexion with the analytical forms called 'trees'. *Proc. London Math. Soc.* **22** (1890–1), 330–346.

PETERSEN, J. Die Theorie der regulären Graphs, *Acta Math.* **15** (1891), 193–220. [10A]

1892

MACMAHON, P. A. The combination of resistances. *Electrician* **28** (1892), 601–602.

ROUSE BALL, W. W. *Mathematical recreations and problems of past and present times.* (Later entitled *Mathematical recreations and essays.*) Macmillan, London, 1892.

1894

BRUNEL, G. Réseaux réguliers. *Proc.-Verb. Soc. Sci. Phys. Nat. Bordeaux* (1894–5), 3–7.

BRUNEL, G. Polymérisation du carbone. *Proc.-Verb. Soc. Sci. Phys. Nat. Bordeaux* (1894–5), 8–10.

DELANNOY, H. A. Sur les arbres géométriques et leur emploi dans la théorie des combinaisons chimiques. *Compt. Rend. Ass. Franç. Avance. Sci.* **23** Pt. 2 (1894), 102–116.

DELANNOY, H. A. [Réponse à question 20]. *Interméd. Math.* **1** (1894), 72–74.

FRIEDEL, C. Question 20. *Interméd. Math.* **1** (1894), 6.

MANSION, P. Question 51. *Interméd. Math.* **1** (1894), 20.

1895

BROCARD, H. [Réponse à question 360]. *Interméd. Math.* **2** (1895), 232–235.

BRUNEL, G. Recherches sur les réseaux. *Mém. Soc. Sci. Phys. Nat. Bordeaux* (4) **5** (1895), 165–215.

BRUNEL, G. [Remarques sur un problème de combinaisons]. *Mém. Soc. Sci. Phys. Nat. Bordeaux* (4) **5** (1895). *Extr. Proc.-Verb.* xiv–xv.

TARRY, G. Le problème des labyrinthes. *Nouv. Ann. Math.* (3) **14** (1895), 187–190. [1D]

1896

BRUNEL, G. Construction d'un réseau donné à l'aide d'un nombre déterminé de traits. *Proc.-Verb. Soc. Sci. Phys. Nat. Bordeaux* (1895–6), 62–65.

DE LA VALLÉE POUSSIN, C. [Problème des quatre couleurs-deuxième réponse]. *Interméd. Math.* **3** (1896), 179–180.

HEFFTER, L. Über Nachbarconfigurationen, Tripelsysteme und metacyklische Gruppen. *Jahresber. Deut. Math.-Ver.* **5** (1896), 67–68.

MASCHKE, H. The representation of finite groups, especially of the rotation groups of the regular bodies of three- and four-dimensional space, by Cayley's color diagrams. *Amer. J. Math.* **18** (1896), 156–194.

WHITE, H. S. Numerically regular reticulations upon surfaces of deficiency higher than 1. *Bull. Amer. Math. Soc.* **3** (1896–7), 116–121.

1897

AHRENS, W. Über das Gleichungssytem einer Kirchhoff'schen galvanischen Stromverzweigung. *Math. Ann.* **49** (1897), 311–324.

HERRMANN, F. Über das Problem die Anzahl der isomeren Paraffine der Formel C_nH_{2n+2} zu bestimmen. *Ber. Deut. Chem. Ges.* **30** (1897), 2423–2426.

LOSANITSCH, S. M. Die isomerie-Arten bei den Homologen der Paraffin-Reihe. *Ber. Deut. Chem. Ges.* **30** (1897), 1917–1926.

LOSANITSCH, S. M. Bemerkungen zu der Hermannschen Mittheilung: Die Anzahl der isomeren Paraffine. *Ber. Deut. Chem. Ges.* **30** (1897), 3059–3060.

WERNICKE, P. [On the solution of the map color problem]. *Bull. Amer. Math. Soc.* **4** (1897–8), 5.

1898

DAVIS, E. W. Note on special regular reticulations. *Bull. Amer. Math. Soc.* **4** (1897–8), 529–530.

FITTING, F. Question 1228. *Interméd. Math.* **5** (1898), 29–30.

GOURSAT, É. [Sur le théorème de Tait]. *Interméd. Math.* **5** (1898), 251.

HEAWOOD, P. J. On the four-colour map theorem. *Quart. J. Pure Appl. Math.* **29** (1898), 270–285.

HEFFTER, L. Über metacyklische Gruppen und Nachbarconfigurationen. *Math. Ann.* **50** (1898), 261–268.

LEMOINE, É [Sur le théorème de Tait]. *Interméd. Math.* **5** (1898), 251.

PETERSEN, J. [Sur le théorème de Tait]. *Interméd. Math.* **5** (1898), 225–227. [10B]

WHITE, H. S. The construction of special regular reticulations on a closed surface. *Bull. Amer. Math. Soc.* **4** (1897–8), 376–382.

1899

BRUNEL, G. Sur la représentation graphique des isomères. *Proc.-Verb. Soc. Sci. Phys. Nat. Bordeaux* (1898–9), 108–110.

BRUNEL, G. [Réponse à question 1467]. *Interméd. Math.* **6** (1899), 209–210.

DE POLIGNAC, C. Sur le théorème de Tait. *Bull. Soc. Math. France* **27** (1899), 142–145.

LEMOINE, É. Question 1467 [Problème des communications]. *Interméd. Math.* **6** (1899), 51–52.

PETERSEN, J. [Réponse à question 360]. *Interméd. Math.* **6** (1899), 36–38.

1900

POINCARÉ, H. Second complément à l'Analysis Situs. *Proc. London Math. Soc.* **32** (1900), 277–308 = *Oeuvres*, Vol. 6, 338–372.

1901

AHRENS, W. *Mathematische Unterhaltungen und Spiele*. Leipzig, 1901.

1902

DIXON, A. C. On map colouring. *Mess. Math.* (New Ser.) **32** (1902–3), 81–83.

FEUSSNER, W. Über Stromverzweigung in Netzförmigen Leitern. *Ann. Phys.* **9** (1902), 1304–1329.

ROBERTS, S. Networks. *Proc. London Math. Soc.* **34** (1901–2), 259–274.

1904

FEUSSNER, W. Zur Berechnung der Stromstärke in Netzförmigen. *Ann. Phys.* **15** (1904), 385–394.

KÖNIG, D. *Matematikai mulatságok*, Vol. 2. Magyar Könyvtár No. 418, Budapest, 1904.

WERNICKE, P. Über den kartographischen Vierfarbensatz. *Math. Ann.* **58** (1904), 413–426.

1905

KÖNIG, D. A térképszinezésről. *Mat. Fiz. Lapok* **14** (1905), 193–200.

WILSON, J. C. *On the traversing of geometrical figures*. Clarendon Press, Oxford, 1905.

1906

WILSON, J. C. On a supposed solution of the 'four-colour problem'. *Math. Gaz.* **3** (1904–6), 338–340.

1907

DEHN, M. and HEEGAARD, P. Analysis situs. *Encyklopädie der mathematischen Wissenschaften* IIIAB3 (1907), 153–220.

HILTON, H. An application of Cayley's colour-groups. *Quart J. Pure Appl. Math.* **38** (1907), 382–384.

MANTEL, W. [Problem 28]. *Wisk. Opg. Wisk. Genoot.* (New Ser.) **10** (1906–10), 60.

WYTHOFF, W. A. [Solution to problem 28]. *Wisk. Opg. Wisk. Genoot.* (New Ser.) **10** (1906–10), 60–61.

1910

TIETZE, H. Einige Bemerkungen über das Problem des Kartenfärbens auf einseitigen Flächen. *Jahresber. Deut. Math.-Ver.* **19** (1910), 155–159. [7C]

1911

KÖNIG, D. Vonalrendszerek kétoldalú felületeken. *Mat. Természettud. Értesítő* **29** (1911), 112–117.

1912

BIRKHOFF, G. D. A determinant formula for the number of ways of coloring a map. *Ann. of Math.* (2) **14** (1912–3), 42–46 = *Math. Papers*, Vol. 3, 1–5. [9B]

VEBLEN, O. An application of modular equations in analysis situs. *Ann. of Math.* (2) **14** (1912–3), 86–94. [9A]

1913

BIRKHOFF, G. D. The reducibility of maps. *Amer. J. Math.* **35** (1913), 115–128 = *Math. Papers*, Vol. 3, 6–19.

1916

KÖNIG, D. Gráfok és alkalmazásuk a determinánsok és halmazok elméletében. *Mat. Természettud. Értesítő* **34** (1916), 104–119 = Über Graphen und ihre Anwendung auf Determinantentheorie und Mengenlehre. *Math. Ann.* **77** (1916), 453–465. [10D]

1917

BRAHANA, H. R. A proof of Petersen's theorem. *Ann. of Math.* (2) **19** (1917–8), 59–63.

KOWALEWSKI, A. W. R. Hamilton's Dodekaederaufgabe als Buntordnungsproblem. *Sitzungsber. Akad. Wiss. Wien* (Abt IIa) **126** (1917), 67–90.

KOWALEWSKI, A. Topologische Deutung von Buntordnungsproblemen. *Sitzungsber. Akad. Wiss. Wien* (Abt IIa) **126** (1917), 963–1007.

SKOLEM, T. Untersuchungen über einige Klassen kombinatorischer Probleme. *Videnskapsselskapets Skrifter* (I. Mat.–Natur. Kl.) No. 6 (1917), 1–99.

1918

PRÜFER, H. Neuer Beweis eines Satzes über Permutationen. *Arch. Math. Phys.* (3) **27** (1918), 142–144. [3D]

1921

ERRERA, A. *Du coloriage des cartes et de quelques questions d'analysis situs.* Falk fils, Brussels, and Gauthier-Villars, Paris, 1921.

1922

CHUARD, J. Questions d'analysis situs. *Rendic. Circ. Mat. Palermo* **46** (1922), 185–224.

ERRERA, A. Une démonstration du théorème de Petersen. *Mathesis* **36** (1922), 56–61.

FRANKLIN, P. The four color problem. *Amer. J. Math.* **44** (1922), 225–236. [9C]

VEBLEN, O. *Analysis situs.* (Amer. Math. Soc. Colloq. Lect. 1916), New York, 1922. [8B]

1923

BRAHANA, H. R. The four-color problem. *Amer. Math. Monthly* **30** (1923), 234–243.

CHUARD, J. Quelques propriétés des réseaux cubiques tracés sur une sphère. *Compt. Rend. Acad. Sci.* (*Paris*) **176** (1923), 73–75.

CHUARD, J. Sur un théorème relatif a certains réseaux cubiques sur une sphère. *Enseign. Math.* **23** (1923), 209.

CHUARD, J. Le problème des quatre couleurs. *Enseign. Math.* **22** (1923), 373–374.

ERRERA, A. Un théorème sur les liaisons. *Compt. Rend. Acad. Sci.* (*Paris*) **177** (1923), 489–491.

ERRERA, A. Le problème des quatre couleurs. *Enseign. Math.* **23** (1923), 95–96.

KÖNIG, D. Sur un problème de la théorie générale des ensembles et la théorie des graphes. *Rev. Métaphys. Morale* **30** (1923), 443–449.

LEVI, F. Streckenkomplexe auf Flächen. *Math. Z.* **16** (1923), 148–158.

SAINTE-LAGUË, A. Les réseaux. *Compt. Rend. Acad. Sci.* (*Paris*) **176** (1923), 1202–1205.

WEYL, H. Repartición de corriente en una red conductora (Introducción al análisis combinatorio). *Rev. Mat. Hispano-Amer.* **5** (1923), 153–164 = *Ges. Abh.*, Vol. 2, 378–389.

1924

ERRERA, A. Sur le problème des quatre couleurs. *Compt. Rend. Ass. Franç. Avance. Sci.* Liège (1924), 96–99.

ERRERA, A. Quelques remarques sur le problème des quatre couleurs. *Proc. Int. Congr. Math.* Toronto, Vol. 1, 1924, 693–694.

REYNOLDS, C. N. Note on the map coloring problem. *Bull. Amer. Math. Soc.* **30** (1924), 220.

1925

ERRERA, A. Une contribution au problème des quatre couleurs. *Bult. Soc. Math. France* **53** (1925), 42–55.

FRANKLIN, P. The electric currents in a network. *J. Math. Phys.* **4** (1925), 97–102.

LEVI, F. Streckenkomplexe auf Flächen (Zweite Mitteilung). *Math. Z.* **22** (1925), 45–61.

REYNOLDS, C. On the map-coloring problem, with particular reference to connected sets of pentagons. *Bull. Amer. Math. Soc.* **31** (1925), 297–298.

SPEISER, A. [Translation of Euler's 1736 paper]. *Klassische Stücke der Mathematik.* Zürich and Leipzig, 1925, 127–138.

1926

FRINK, O. A proof of Petersen's theorem. *Ann. of Math.* (2) **27** (1925–6), 491–493. [10C]

KÖNIG, D. and VALKÓ, S. Halmazok többértelmű leképzéséről. *Mat. Természettud. Értesítő* **42** (1926), 173–176 = Über mehrdeutige Abbildungen von Mengen. *Math. Ann.* **95** (1926), 135–138.

REYNOLDS, C. N. On the problem of coloring maps in four colors I. *Ann. of Math.* (2) **28** (1926–7), 1–15.

REYNOLDS, C. N. On the problem of coloring maps in four colors. *Compt. Rend. Ass. Franç. Avance. Sci.* Lyon (1926), 88–91.

SAINTE-LAGUË, A. Les réseaux unicursaux et bicursaux. *Compt. Rend. Acad. Sci. (Paris)* **182** (1926), 747–750.

SAINTE-LAGUË, A. Les réseaux (ou graphes). Mémorial des Sciences Math. fasc. 18, Gauthier-Villars, Paris, 1926.

1927

ERRERA, A. Exposé historique du problème des quatre couleurs. *Period. Mat.* (4) **7** (1927), 20–41.

KÖNIG, D. Über eine Schlussweise aus dem Endlichen ins Unendliche (Punktmengen— Kartenfärben—Verwandtschaftsbeziehungen—Schachspiel). *Acta. Litt. Sci. Szeged* **3** (1927), 121–130.

MENGER, K. Zur allgemeinen Kurventheorie. *Fund. Math.* **10** (1927), 96–115.

REDFIELD, J. H. The theory of group-reduced distributions. *Amer. J. Math.* **49** (1927), 433–455.

REYNOLDS, C. N. On the problem of coloring maps in four colors II. *Ann. of Math.* (2) **28** (1926–7), 477–492.

SKOLEM, T. Chap. 15, 292–308 of: Netto, E. *Lehrbuch der Combinatorik.* 2nd ed., Teubner, Stuttgart, 1927. Reprinted: Chelsea, New York, 1958.

1928

GONSETH, F. and JUVET, G. Sur le problème des quatre couleurs. *Atti. Congr. Int. Math.* Bologna, Vol. 4, 1928, 363–365.

1929

LUNN, A. C. and SENIOR, J. K. Isomerism and configuration. *J. Phys. Chem.* **33** (1929), 1027–1079.

SAINTE-LAGUË, A. Géométrie de situation et jeux. Mémorial des Science Math., fasc. 41. Gauthier-Villars, Paris, 1929.

1930

BIRKHOFF, G. D. On the number of ways of colouring a map. *Proc. Edinb. Math. Soc.* (2) **2** (1930), 83–91 = *Math. Papers*, Vol. 3, 20–28.

KURATOWSKI, K. Sur le problème des courbes gauches en topologie. *Fund. Math.* **15** (1930), 271–283. [8C]

MENGER, K. Über plättbare Dreiergraphen und Potenzen nichtplättbarer Graphen. *Anz. Akad. Wiss. Wien* **67** (1930), 85–86 = *Ergebnisse eines Math. Kolloquiums*, Vol. 2, 1930, 30–31.

RAMSEY, F. P. On a problem of formal logic. *Proc. London Math. Soc.* (2) **30** (1930), 264–286.

1931

BAKER, R. P. Cayley diagrams on the anchor ring. *Amer. J. Math.* **53** (1931), 645–669.

KÖNIG, D. Graphok és matrixok. *Mat. Fiz. Lapok* **38** (1931), 116–119.

WHITNEY, H. A theorem on graphs. *Ann. of Math.* (2) **32** (1931), 378–390.

WHITNEY, H. The coloring of graphs. *Proc. Nat. Acad. Sci. U.S.A.* **17** (1931), 122–125.

WHITNEY, H. Non-separable and planar graphs. *Proc. Nat. Acad. Sci. U.S.A.* **17** (1931), 125–127.

1932

CHUARD, J. Les réseaux cubiques et le problème des quatre couleurs. *Mém. Soc. Vaudoise Sci. Nat.* **4** (1932), 41–101.

CHUARD, J. Une solution du problème des quatre couleurs. *Verhandl. Int. Math. Kongr.* Zürich, Vol. 2, 1932, 199–200.

FOSTER, R. M. Geometrical circuits of electrical networks. *Trans. Amer. Inst. Elec. Eng.* **51** (1932), 309–317.

HEAWOOD, P. J. On extended congruences connected with the four-colour map theorem. *Proc. London Math. Soc.* (2) **33** (1932), 253–286.

KRAHN, E. Die Wahrscheinlichkeit der Richtigkeit des Vierfarbensatzes. *Acta Comment. Univ. Tartu* (A) **22**, No. 2 (1932), 1–7.

PANNWITZ, E. [Review of Chuard's paper]. *Jahrb. Fortschr. Math.* **58** (1932), 1204.

REYNOLDS, C. N. Circuits upon polyhedra. *Ann of Math.* (2) **33** (1932), 367–372.

STUCKE, E. Ein mathematisch-ästhetisches Problem. *Z. für Math. Wiss. Unterricht* **63** (1932), 110–112.

WHITNEY, H. Congruent graphs and the connectivity of graphs. *Amer. J. Math.* **54** (1932), 150–168.

WHITNEY, H. The coloring of graphs. *Ann. of Math.* (2) **33** (1932), 688–718.

WHITNEY, H. [Conditions that a graph have a dual]. *Bull. Amer. Math. Soc.* **38** (1932), 37.

WHITNEY, H. A logical expansion in mathematics. *Bull. Amer. Math. Soc.* **38** (1932), 572–579. [9D]

WHITNEY, H. Non-separable and planar graphs. *Trans. Amer. Math. Soc.* **34** (1932), 339–362. [8D]

1933

DIXON, A. C. The specification of a map. *J. London Math. Soc.* **8** (1933), 134–136.

KÖNIG, D. Über trennende Knotenpunkte in Graphen. *Acta Litt. Sci. Szeged* **6** (1933), 155–179.

WHITNEY, H. A set of topological invariants for graphs. *Amer. J. Math.* **55** (1933), 231–235.

WHITNEY, H. On the classification of graphs. *Amer. J. Math.* **55** (1933), 236–244.

WHITNEY, H. 2-isomorphic graphs. *Amer. J. Math.* **55** (1933), 245–254.

WHITNEY, H. Planar graphs. *Fund. Math.* **21** (1933), 73–84.

WHITNEY, H. A characterisation of the closed 2-cell. *Trans. Amer. Math. Soc.* **35** (1933), 261–273.

ZYLINSKI, E. Deux remarques sur les réseaux. *Bull. Int. Acad. Pol. Sci.* (A) No. 9 (1933), 295–297.

1934

BIRKHOFF, G. D. On the polynomial expressions for the number of ways of coloring a map. *Ann. Scuola Norm. Sup. Pisa* (2) **3** (1934), 85–104 = *Math. Papers*, Vol. 3, 29–47.

FRANKLIN, P. A six-color problem. *J. Math. Phys.* **13** (1934), 363–369.

HAJÓS, G. Zum Mengerschen Graphensatz. *Acta Litt. Sci. Szeged* **7** (1934), 44–47.

RÉDEI, L. Ein kombinatorischer Satz. *Acta Litt. Sci. Szeged* **7** (1934), 39–43.

SCHÖNBERGER, T. Ein Beweis des Petersenschen Graphensatzes. *Acta Litt. Sci. Szeged* **7** (1934), 51–57.

1935

HALL, P. On representatives of subsets. *J. London Math. Soc.* **10** (1935), 26–30.

KAGNO, I. N. A note on the Heawood color formula. *J. Math. Phys.* **14** (1935), 228–231.

KITTELL, I. A group of operations on a partially colored map. *Bull. Amer. Math. Soc.* **41** (1935), 407–413.

MACLANE, S. Some unique separation theorems for graphs. *Amer. J. Math.* **57** (1935), 805–820.

PÓLYA, G. Un problème combinatoire général sur les groupes de permutations et le calcul du nombre des isomères des composés organiques. *Compt. Rend. Acad. Sci. (Paris)* **201** (1935), 1167–1169. [4D]

WHITNEY, H. On the abstract properties of linear dependence. *Amer. J. Math.* **57** (1935), 509–533.

1936

ERDŐS, P., GRÜNWALD, T., and WEISZFELD, E. Végtelen gráfok Euler-vonalairól. *Mat. Fiz. Lapok* **43** (1936), 129–141.

HARMEGNIES, R. Sur quelques propriétés des reséaux. *Compt. Rend. Acad. Sci. (Paris)* **202** (1936), 1142–1144.

HEAWOOD, P. J. Failures in congruences connected with the four-colour map theorem. *Proc. London Math. Soc.* (2) **40** (1936), 189–202.

HEEGAARD, P. Bemerkungen zum Vierfarbenproblem. *Mat. Sb.* (New Ser.) **1** (1936), 685–693.

KAGNO, I. N. The triangulation of surfaces and the Heawood color formula. *J. Math. Phys.* **15** (1936), 179–186.

KÖNIG, D. *Theorie der endlichen und unendlichen Graphen.* Akademische Verlagsgesellschaft, Leipzig, 1936. Reprinted: Chelsea, New York, 1950.

MOTZKIN, T. Contributions à la théorie des graphes. *Compt. Rend. Congr. Int. Math.* Oslo, Vol. 2, 1936, 133–134.

PÓLYA, G. Kombinatorische Anzahlbestimmungen für Permutationsgruppen und chemische Verbindungen. *Compt. Rend. Congr. Int. Math.* Oslo, Vol. 2, 1936, 19.

PÓLYA, G. Sur le nombre des isomères de certains composés chimiques. *Compt. Rend. Acad. Sci. (Paris)* **202** (1936), 1554–1556.

PÓLYA, G. Algebraische Berechnung der Anzahl der Isomeren einiger organischer Verbindungen. *Z. Kristal.* (A) **93** (1936), 415–443.

RATIB, I. and WINN, C. E. Généralisation d'une réduction d'Errera dans le problème des quatre couleurs. *Compt. Rend. Congr. Int. Math.* Oslo, Vol. 2, 1936, 131–133.

WAGNER, K. Bemerkung zum Vierfarbenproblem. *Jahresber. Deut. Math.-Verein.* **46** Pt. 1, (1936), 26–32.

WAGNER, K. Ein Satz über Komplexe. *Jahresber. Deut. Math.-Verein.* **46** Pt. 2 (1936), 21–22.

WAGNER, K. Zwei Bemerkungen über Komplexe. *Math. Ann.* **112** (1936), 316–321.

Index of Names

Capital letters are used for those whose biographies are given in Appendix 2; a number with an asterisk refers to a portrait of the person concerned.

General Index

This index gives the main reference for each of the terms included; when a given term features in an important way in more than one section of the book, a reference is given for each occurrence.